LONDON MATHEMATICAL SOCIETY LECTURE NOTE SERIES

Managing Editor: Professor N.J. Hitchin, Mathematical Institute,
University of Oxford, 24–29 St. Giles, Oxford OX1 3LB, United Kingdom

The titles below are available from booksellers, or, from Cambridge University Press at www.cambridge.org

London Mathematical Society Lecture Note Series. 334

The Navier–Stokes equations: a classification of flows and exact solutions

P. G. DRAZIN
University of Bristol

N. RILEY
University of East Anglia

CAMBRIDGE UNIVERSITY PRESS

CAMBRIDGE UNIVERSITY PRESS
Cambridge, New York, Melbourne, Madrid, Cape Town,
Singapore, São Paulo, Delhi, Mexico City

Cambridge University Press
The Edinburgh Building, Cambridge CB2 8RU, UK

Published in the United States of America by Cambridge University Press, New York

www.cambridge.org
Information on this title: www.cambridge.org/9780521681629

First published 2006
Reprinted 2007

A catalogue record for this publication is available from the British Library

Library of Congress Cataloguing in Publication Data

ISBN 978-0-521-68162-9 Paperback

Contents

Preface

The origins of this book date to a conversation between the authors a short time before both were due to (formally) retire. Sadly little had been achieved before the more experienced author died. As a consequence any shortcomings of the book must be attributed to the surviving author.

Exact solutions of any system of partial differential equations attract attention. This must be particularly true of the Navier–Stokes equations which, for the best part of 200 years, have been the foundation for the significant and world-wide study of the behaviour of fluids in motion. The subject burgeoned in the twentieth century from stimuli as diverse as international conflict, and a desire to create a better understanding of the environment. In the nineteenth century theoretical advance was slow, and until the approximate or, as we would rather view them, asymptotic theories of Stokes and Prandtl for small and large values of the Reynolds number were devised, only exact solutions, and few at that, were available. In spite of the advances in asymptotic methods during the first half of the twentieth century, and the increasing use of computational methods in its later decades, exact solutions of the Navier–Stokes equations have been pursued. At best these provide an insight into the behaviour of fluids in motion; they may also provide a vehicle for novel mathematical methods or a useful check for a computer code. Some, it must be admitted, provide little of value in either of these senses. The end of the twentieth century appeared to the authors, when computational methods were clearly dominating theoretical research in fluid dynamics, to be an appropriate time to survey the range of exact solutions. We have brought together material from diverse sources, and attempted to present that material in a uniform and coherent manner. The monograph may be viewed as a supplement or complement to the earlier extensive review by R. Berker, and subsequent review articles by C. Y. Wang. We believe that the content will be of value to those with an interest in the subject of fluid mechanics, and not least to those who are teaching or learning the subject.

In dedicating the book to the memories of Leslie Howarth, James Lighthill and Keith Stewartson we are mindful of their significant contributions, not only to fluid mechanics but also to the careers of the authors.

N. Riley
Norwich 2005

1

Scope of the book

This monograph elaborates a fundamental topic of the theory of fluid dynamics which is introduced in most textbooks on the theory of flow of a viscous fluid. A knowledge of this introductory background, for which reference may be made to Batchelor (1967), will be assumed here. However, it will be helpful to summarise a little of the background wherever we need it. In particular, we begin by introducing the scope of the book by loosely defining the terms of the title.

The *Navier–Stokes equations* are the system of non-linear partial differential equations governing the motion of a *Newtonian fluid*, which may be liquid or gas. In essence, they represent the balance between the rate of change of momentum of an element of fluid and the forces on it, as does Newton's second law of motion for a particle, where the stress is linearly related to the rate of strain of the fluid. Newton himself did not understand well the nature of the forces between elemental particles in a continuum, but he did (Newton 1687, Vol. II, Section IX, *Hypothesis, Proposition LI*) initiate the theory of the dynamics of a uniform viscous fluid in an intuitive and imaginative way. It was many years later that the Navier–Stokes equations, as we now know them, were deduced from various physical hypotheses, and in various forms, by Navier (1827), Poisson (1831), Saint-Venant (1843) and Stokes (1845). Stokes (1846) himself reviewed the methods and hypotheses of these authors, and presented a short rational derivation of the equations. We present a summary here; for a formal derivation the reader is referred to modern treatments by Batchelor (1967), Long (1961a), and Whitham (1963).

The equations may be expressed in the form

$$\rho \frac{\mathrm{D}\mathbf{v}}{\mathrm{D}t} = \rho \mathbf{F} + \nabla.\sigma \tag{1.1}$$

where $\rho(\mathbf{x}, t)$ is the density, $\mathbf{v}(\mathbf{x}, t)$ the velocity vector, $\mathbf{F}(\mathbf{x}, t)$ the body-force vector per unit mass, and $\boldsymbol{\sigma}$ the stress tensor of the fluid at a point with position vector $\mathbf{x}$ at time t; the *material* or *convective* derivative of a field varying with $\mathbf{x}$ and t is defined as

$$\frac{D}{Dt} = \frac{\partial}{\partial t} + \mathbf{v}.\nabla. \tag{1.2}$$

For a Newtonian fluid the stress tensor is a linear function of the rate of strain, from which it may be deduced that its components are

$$\sigma_{ij} = -p\delta_{ij} + \lambda \frac{\partial v_k}{\partial x_k}\delta_{ij} + \mu \left(\frac{\partial v_i}{\partial x_j} + \frac{\partial v_j}{\partial x_i} \right), \tag{1.3}$$

where λ is the second coefficient of viscosity and μ the coefficient of viscosity of the fluid. In general λ and μ depend upon the thermodynamic properties of the fluid, and may vary in space and time. However, we shall assume they are uniform in space and constant in time.

These equations have been widely accepted as an excellent model of the macroscopic motions of most real fluids, including air and water, and are used by countless engineers, physicists, chemists, mathematicians, meteorologists, oceanographers, geologists and biologists. There are, however, notable and useful models of fluids whose motions are not governed by the Navier–Stokes equations. For example there are *non-Newtonian fluids* which are governed by a non-linear stress tensor, and *visco-elastic* fluids in which the stress depends on the strain as well as on the rate of strain of the fluid and retains a 'memory' of previous deformations; see Spencer (1980).

In addition to conservation of momentum (1.1) we have conservation of mass which leads to the well known continuity equation

$$\frac{1}{\rho}\frac{D\rho}{Dt} + \nabla.\mathbf{v} = 0. \tag{1.4}$$

In addition to the above equations the body force $\mathbf{F}$ must be specified, together with an equation of state, if we are to model the motion of the fluid. For the example of a uniform incompressible fluid, which we adopt throughout, the equation of state is simply that the density is constant, so the equation of continuity becomes

$$\nabla.\mathbf{v} = 0. \tag{1.5}$$

The implication of (1.5) is that the stress (1.3) is independent of λ.

For any given problem in fluid dynamics, boundary conditions and initial conditions must be specified. Whilst such conditions are specific to the problem under consideration, it should be noted that an exact solution typically arises with certain boundary conditions sharing the same symmetries as assumed in deriving the solution. In other words conditions specific to a problem do not in general coincide with those necessary to find an exact solution. At a solid boundary it is now generally accepted that there is no relative motion between the boundary and the fluid. This not only implies non-permeability, but also that there is no slip between the fluid and solid. The no-slip condition proved controversial throughout the nineteenth century; Goldstein (1938, pp. 676–680) has given an account of the long history of the controversy. At an interface between two immiscible fluids the velocity and stress components must be continuous; at an air–liquid interface, as in water-wave theory, it is common practice to approximate these conditions by vanishing shear stress, and constant pressure.

An alternative form of equation (1.1) is

$$\frac{\partial \mathbf{v}}{\partial t} - \mathbf{v} \wedge \boldsymbol{\omega} = -\frac{1}{\rho} \nabla \left(p + \frac{1}{2}\rho v^2 \right) + \mathbf{F} - \nu \nabla \wedge \boldsymbol{\omega}, \tag{1.6}$$

where $\boldsymbol{\omega} = \nabla \wedge \mathbf{v}$ is the vorticity vector and $\nu = \mu/\rho$ is the kinematic viscosity. Physically, the vorticity at any point in a fluid flow is proportional to the instantaneous angular velocity of an elementary spherical particle of fluid centred on the point. Flows with $\boldsymbol{\omega} \equiv \mathbf{0}$ are known as irrotational flows for which, of course, $\mathbf{v} = \nabla \phi$. Taking the 'curl' of equation (1.6), and acknowledging the solenoidal property of $\boldsymbol{\omega}$, gives

$$\frac{D\boldsymbol{\omega}}{Dt} = (\boldsymbol{\omega}.\nabla)\mathbf{v} + \nabla \wedge \mathbf{F} + \nu \nabla^2 \boldsymbol{\omega}, \tag{1.7}$$

where the terms on the right-hand side of (1.7) represent contributions to the rate of change of vorticity at a fluid particle. Thus, the contribution of the first term is due to the stretching and bending of vortex lines, the second is a distributed source term that is only relevant when the body force is non-conservative, whilst the last term represents the diffusion of vorticity between fluid particles. In general we shall assume that the body force is zero. There is a sense in which the vorticity may be viewed as fundamental, since it is the only flow quantity whose values are not propagated instantaneously in an incompressible fluid.

For an inviscid fluid, Lagrange's theorem tells us that a fluid which initially has zero vorticity, will have vanishing vorticity for all time. This cannot be true in a viscous fluid, since even in the absence of non-conservative body forces diffusion of vorticity from nearby particles can occur. But diffusion alone cannot *create* vorticity, and a question we must ask is how vorticity is introduced into the fluid where none is present initially? The answer is that a solid boundary acts as a source of vorticity. To appreciate this consider a two-dimensional flow, under the action of a pressure gradient, in the x-direction over the plane boundary $y = 0$. With $\mathbf{v} = (u, v, 0)$, $\boldsymbol{\omega} = (0, 0, \zeta)$ we have the flow of vorticity out of the surface $y = 0$, using (1.6) with $\mathbf{F} = \mathbf{0}$, as

$$-\nu\frac{\partial \zeta}{\partial y} = -\nu\frac{\partial}{\partial y}\left(\frac{\partial v}{\partial x} - \frac{\partial u}{\partial y}\right) = \nu\frac{\partial^2 u}{\partial y^2} = \frac{1}{\rho}\frac{\partial p}{\partial x}.$$

We see that (negative) vorticity flows into the fluid from the boundary in a favourable pressure gradient $\partial p/\partial x < 0$; when this changes sign vorticity is removed from it at the surface. This may rapidly reduce the magnitude of the surface vorticity to zero, corresponding to separation of the flow from the boundary. We shall often find it convenient to interpret our exact solutions in terms of their vortex dynamics.

An *exact solution* may seem to be no more nor less than a solution, because either a given set of fields $\mathbf{v}$, p, for given ρ, μ and body force $\mathbf{F}$ satisfies the governing equations or it does not. However, usage has given the phrase 'exact solution' a special meaning. It often denotes a solution which has a simple explicit form, usually an expression in finite terms of elementary or other well known special functions. Sometimes an exact solution is taken to be one which can be reduced to the solution of an ordinary differential equation or a system of a few ordinary differential equations. Rarely we go even further, and take an exact solution to be the solution of a partial differential equation, provided that the equation has fewer independent variables than the Navier–Stokes equations themselves. This is in contrast to an 'approximate solution', which is taken to be a field, simple or complicated, which approximates a solution either in a numerical sense or in an asymptotic limit, for example vanishingly small viscosity. Thus the logical distinctions between solutions, exact solutions and approximate solutions are blurred, but in practice the distinctions made are usually clear and useful. The exact solutions are, essentially, a subset of the solutions of the Navier–Stokes equations which happen to have relatively simple mathematical expressions and which are, mostly, simple physically. The essence of this account, then, is the explicitness and relative simplicity of the expression of the solutions. Many exact solutions of the Navier–Stokes equations are unstable and therefore unobservable in practice. The stability of flows is a large and important topic of fluid mechanics in its own right. We choose not to introduce it in this work, but refer the reader to two modern accounts of the subject, by Drazin and Reid (1981), and Drazin (2002).

In the early decades of the development of the mathematical theory of the motion of a viscous fluid, exact solutions were the only solutions available. Researchers solved what problems they could, rather than solving the practical problems in hand. Inevitably the solvable problems were the simple ones, usually idealised with a strong symmetry. From the mid-nineteenth century, and early twentieth century, asymptotic methods were developed, and thereafter numerical methods, albeit crude ones at first, to extend somewhat the range of tractable problems. However, the development of computational fluid dynamics over the past half century has changed the emphasis of research from, as one might express it, finding exact solutions of approximate problems to finding approximate solutions of exact problems. Nevertheless, the exact solutions remain a valuable and irreplaceable resource. They immediately convey more physical insight than a numerical table or sheaf of diagrams. This is especially true when the system is governed by one or more parameters, when a complete numerical tabulation would be voluminous. It may be argued that the exact solutions convey more than does the watching of a video of a numerical experiment or attendance at an experiment in a laboratory. However, the visual and aural appreciation of laboratory experiments and natural phenomena must never be underestimated. The very simplicity of exact solutions, in many cases, causes them to be used as prototypes which form our concepts of flow; for example when looking at a flow in a channel whose walls are nearly parallel we interpret observations of the flow by reference to our knowledge of plane Poiseuille flow, with a parabolic velocity profile, in a channel with fixed parallel walls. The exact solutions are first approximations to more general flows in many asymptotic theories as well as in our concepts. The exact solutions are also valuable when, as special cases of more general results, they enable theories or computer programs giving those results to be tested for errors and accuracy; in particular, they may provide benchmark tests for computational fluid dynamics. Exact solutions often arise as similarity solutions associated with systems of ordinary non-linear differential equations; many of these have provided a rich source of challenging and fruitful analytical problems and have attracted the interest of mathematical analysts. Such similarity solutions may emerge as asymptotic limits of more general solutions after a long time, or as the far-field solution in some spatial limit. Finally we remark that exact solutions are invaluably useful for examples and exercises that help students to learn and understand the theory of fluid mechanics. In teaching they can provide a bridge to the use of pictures of laboratory and computer experiments and the use of difficult mathematical methods.

We note that the Navier–Stokes equations are invariant under the action of various discrete and continuous groups of transformations of the independent and dependent variables. The set of all solutions of the equations is invariant under the same groups, although an individual solution may change under the action of one or more of the groups, and a given set of boundary conditions and initial conditions may change similarly. The groups of transformations under which the Navier–Stokes equations are invariant are: translations of time and space, Galilean transformations, parity or reversal and the group of transformations $t, \mathbf{x}, \mathbf{v}, p \longmapsto k^2 t, k\mathbf{x}, k^{-1}\mathbf{v}, k^{-2}p$. The groups of transformations due to scalings of time and space are closely related to the ideas of dynamical similarity and Buckingham's pi theorem. If we define $t' = t/T, \mathbf{x}' = \mathbf{x}/L, \mathbf{v}' = \mathbf{v}/U, p' = p/\rho U^2$ for arbitrary scales of time T, length L and velocity U then the Navier–Stokes equations for a uniform incompressible fluid in the absence of a body force are transformed to the dimensionless form

$$St \frac{\partial \mathbf{v}'}{\partial t} + \mathbf{v}'.\nabla'\mathbf{v}' = -\nabla' p' + R^{-1}\nabla'^2\mathbf{v}', \quad \nabla'.\mathbf{v}' = 0, \tag{1.8}$$

where $St = L/UT$ is the Strouhal number and $R = UL/\nu$ is the Reynolds number. For flows in which there is a natural time scale ω^{-1} where, for example, ω is a characteristic frequency, then we see from (1.8) that our unsteady flow is characterised by two independent dimensionless parameters. However, for an unsteady flow, for example a flow started from rest, in which the only time scale is L/U then, as for steady flows, a single parameter R characterises the flow.

Invariance under groups of scaling transformations leads to many exact solutions of the Navier–Stokes equations, known as similarity solutions, which arise as solutions of ordinary differential equations. Whilst most books on viscous flow theory pay some attention to dynamical similarity and similarity solutions, there are some books, for example Sedov (1959), Birkhoff (1960) and Barenblatt (1979, 1996), devoted to the theory of dynamical similarity and scaling transformations in similarity solutions. In particular Barenblatt (1996) gives a stimulating assessment of the significance and importance of the exact solutions of partial differential equations, as well as their relationship to asymptotic properties of other solutions. There are a few books, for example Bluman and Kumei (1989), Olver (1992) and Fushschich, Shtelen and Serov (1993), on the systematic use of continuous groups to find solutions of linear and non-linear partial differential equations, where it can be seen that Lie's theory has been used to find some valuable new solutions of equations other than the Navier–Stokes equations. Nevertheless the maturity of both the theory of the Navier–Stokes equations and the Lie theory of differential equations makes it unlikely that many more exact solutions of the Navier–Stokes equations remain to be discovered. For the same reason, those that do remain are probably of little importance. However, recent developments by Ludlow, Clarkson and Bassom (1998, 1999) in which non-classical reduction methods are employed, as opposed to the classical Lie group method, may offer an alternative way ahead.

Much of the early history of the exact solutions has been recorded by Truesdell (1954); also many textbooks, such as Batchelor (1967), contain brief accounts of the most important exact solutions. More extensive accounts are to be found in treatises on the flow of a viscous fluid, such as Dryden, Murnaghan and Bateman (1932), Schlichting (1979), Whitham (1963) and Lagerstrom (1996). Berker (1963) described the exact solutions in great detail; indeed at the time it was written his article was comprehensive and encyclopaedic. Wang (1989a, 1990a, 1991) has recently written review articles. Yet we believe these works leave a need for an up-to-date comprehensive integrated account of the exact solutions, and that this book fulfils that need. We do not claim that our account is exhaustive, and indeed references published before Berker's (1963) article are emphasised less than those published afterwards. Our philosophy for selection is that a solution is 'significant', either in an historical or novel sense or, perhaps most importantly, that it offers insight into the dynamical behaviour of viscous fluids. In that sense omissions are due to prejudice, but we hope not ignorance. The remaining four chapters of the book divide the exact solutions as follows. In chapter 2 we consider steady flows bounded by plane boundaries, and in chapter 3 steady flows bounded by curved boundaries, or exhibiting axisymmetry. To some extent that division between the two chapters is not clear-cut, since some exact solutions could feature in either. The remaining two chapters are devoted to unsteady flows, with a division as between the first two.

We conclude this introductory chapter by setting out, for subsequent reference, the Navier–Stokes equations for an incompressible fluid, in the three most commonly occurring co-ordinate systems. In what follows we make extensive reference to these. For general orthogonal curvilinear co-ordinates reference may be made to Whitham (1963).

(i) Cartesian co-ordinates In Cartesian co-ordinates (x, y, z) we take respective velocity components (u, v, w) so that $\mathbf{v} = u\mathbf{i} + v\mathbf{j} + w\mathbf{k}$. The Navier–Stokes and continuity equations (1.1), (1.5) then become

$$\frac{\partial u}{\partial t} + u\frac{\partial u}{\partial x} + v\frac{\partial u}{\partial y} + w\frac{\partial u}{\partial z} = -\frac{1}{\rho}\frac{\partial p}{\partial x} + X + v\nabla^2 u, \qquad (1.9)$$

$$\frac{\partial v}{\partial t} + u\frac{\partial v}{\partial x} + v\frac{\partial v}{\partial y} + w\frac{\partial v}{\partial z} = -\frac{1}{\rho}\frac{\partial p}{\partial y} + Y + v\nabla^2 v, \qquad (1.10)$$

$$\frac{\partial w}{\partial t} + u\frac{\partial w}{\partial x} + v\frac{\partial w}{\partial y} + w\frac{\partial w}{\partial z} = -\frac{1}{\rho}\frac{\partial p}{\partial z} + Z + v\nabla^2 w, \qquad (1.11)$$

$$\frac{\partial u}{\partial x} + \frac{\partial v}{\partial y} + \frac{\partial w}{\partial z} = 0, \qquad (1.12)$$

where the body force per unit mass $\mathbf{F} = (X, Y, Z)$ and here, and below, ∇^2 represents the three-dimensional Laplacian operator. The vorticity $\boldsymbol{\omega} = \xi\mathbf{i} + \eta\mathbf{j} + \zeta\mathbf{k} = (\xi, \eta, \zeta)$ where

$$\xi = \frac{\partial w}{\partial y} - \frac{\partial v}{\partial z}, \quad \eta = \frac{\partial u}{\partial z} - \frac{\partial w}{\partial x}, \quad \zeta = \frac{\partial v}{\partial x} - \frac{\partial u}{\partial y}. \tag{1.13}$$

For two-dimensional flow, independent of the co-ordinate z, say, the continuity equation (1.12) is satisfied by introducing the stream function ψ such that

$$u = \frac{\partial \psi}{\partial y}, \quad v = -\frac{\partial \psi}{\partial x}, \tag{1.14}$$

and the only non-zero component of the vorticity is

$$\zeta = -\nabla^2 \psi, \tag{1.15}$$

which satisfies, for a conservative body force $\mathbf{F}$,

$$\frac{D\zeta}{Dt} = \nu\nabla^2\zeta, \tag{1.16}$$

so that

$$\frac{\partial}{\partial t}(\nabla^2\psi) - \frac{\partial(\psi, \nabla^2\psi)}{\partial(x, y)} = \nu\nabla^4\psi. \tag{1.17}$$

(ii) Cylindrical polar co-ordinates We define cylindrical polar co-ordinates (r, θ, z) such that

$$x = r\cos\theta, \quad y = r\sin\theta, \quad r \geq 0, \quad 0 \leq \theta < 2\pi, \tag{1.18}$$

with corresponding velocity components $\mathbf{v} = (v_r, v_\theta, v_z) = v_r\hat{\mathbf{r}} + v_\theta\hat{\boldsymbol{\theta}} + v_z\hat{\mathbf{z}}$, vorticity and body-force components $\boldsymbol{\omega} = (\omega_r, \omega_\theta, \omega_z)$, $\mathbf{F} = (F_r, F_\theta, F_z)$ respectively. The components of the Navier–Stokes equations are, then,

$$\frac{\partial v_r}{\partial t} + v_r\frac{\partial v_r}{\partial r} + \frac{v_\theta}{r}\frac{\partial v_r}{\partial \theta} + v_z\frac{\partial v_r}{\partial z} - \frac{v_\theta^2}{r}$$
$$= -\frac{1}{\rho}\frac{\partial p}{\partial r} + F_r + \nu\left(\nabla^2 v_r - \frac{v_r}{r^2} - \frac{2}{r^2}\frac{\partial v_\theta}{\partial \theta}\right), \tag{1.19}$$

$$\frac{\partial v_\theta}{\partial t} + v_r\frac{\partial v_\theta}{\partial r} + \frac{v_\theta}{r}\frac{\partial v_\theta}{\partial \theta} + v_z\frac{\partial v_\theta}{\partial z} + \frac{v_r v_\theta}{r}$$
$$= -\frac{1}{\rho}\frac{1}{r}\frac{\partial p}{\partial \theta} + F_\theta + \nu\left(\nabla^2 v_\theta + \frac{2}{r^2}\frac{\partial v_r}{\partial \theta} - \frac{v_\theta}{r^2}\right), \tag{1.20}$$

$$\frac{\partial v_z}{\partial t} + v_r\frac{\partial v_z}{\partial r} + \frac{v_\theta}{r}\frac{\partial v_z}{\partial \theta} + v_z\frac{\partial v_z}{\partial z} = -\frac{1}{\rho}\frac{\partial p}{\partial z} + F_z + \nu\nabla^2 v_z, \tag{1.21}$$

with the continuity equation

$$\frac{1}{r}\frac{\partial}{\partial r}(rv_r) + \frac{1}{r}\frac{\partial v_\theta}{\partial \theta} + \frac{\partial v_z}{\partial z} = 0. \tag{1.22}$$

The components of vorticity are given by

$$\omega_r = \frac{1}{r}\frac{\partial v_z}{\partial \theta} - \frac{\partial v_\theta}{\partial z}, \qquad \omega_\theta = \frac{\partial v_r}{\partial z} - \frac{\partial v_z}{\partial r}, \qquad \omega_z = \frac{1}{r}\frac{\partial}{\partial r}(rv_\theta) - \frac{1}{r}\frac{\partial v_r}{\partial \theta}. \tag{1.23}$$

For a rotationally symmetric flow, independent of θ, we introduce a different stream function ψ, first identified by Stokes (1842) such that with

$$v_r = -\frac{1}{r}\frac{\partial \psi}{\partial z}, \qquad v_z = \frac{1}{r}\frac{\partial \psi}{\partial r}, \tag{1.24}$$

the continuity equation is satisfied identically.

(iii) Spherical polar co-ordinates We define spherical polar co-ordinates (r, θ, ϕ) such that

$$x = r\sin\theta\cos\phi, \quad y = r\sin\theta\sin\phi, \quad z = r\cos\theta, \quad r \geq 0, \quad 0 \leq \theta \leq \pi,$$

$$0 \leq \phi < 2\pi,$$

with corresponding velocity components $\mathbf{v} = (v_r, v_\theta, v_\phi) = v_r\hat{\mathbf{r}} + v_\theta\hat{\boldsymbol{\theta}} + v_\phi\hat{\boldsymbol{\phi}}$, vorticity and body-force components $\boldsymbol{\omega} = (\omega_r, \omega_\theta, \omega_\phi)$, $\mathbf{F} = (F_r, F_\theta, F_\phi)$ respectively. The components of the Navier–Stokes equations are, then,

$$\frac{\partial v_r}{\partial t} + v_r\frac{\partial v_r}{\partial r} + \frac{v_\theta}{r}\frac{\partial v_r}{\partial \theta} + \frac{v_\phi}{r\sin\theta}\frac{\partial v_r}{\partial \phi} - \frac{v_\theta^2 + v_\phi^2}{r}$$
$$= -\frac{1}{\rho}\frac{\partial p}{\partial r} + F_r + \nu\left(\nabla^2 v_r - \frac{2v_r}{r^2} - \frac{2}{r^2}\frac{\partial v_\theta}{\partial \theta} - \frac{2v_\theta\cot\theta}{r^2} - \frac{2}{r^2\sin\theta}\frac{\partial v_\phi}{\partial \phi}\right), \tag{1.25}$$

$$\frac{\partial v_\theta}{\partial t} + v_r\frac{\partial v_\theta}{\partial r} + \frac{v_\theta}{r}\frac{\partial v_\theta}{\partial \theta} + \frac{v_\phi}{r\sin\theta}\frac{\partial v_\theta}{\partial \phi} + \frac{v_r v_\theta}{r} - \frac{v_\phi^2\cot\theta}{r}$$
$$= -\frac{1}{\rho}\frac{1}{r}\frac{\partial p}{\partial \theta} + F_\theta + \nu\left(\nabla^2 v_\theta + \frac{2}{r^2}\frac{\partial v_r}{\partial \theta} - \frac{v_\theta}{r^2\sin^2\theta} - \frac{2\cos\theta}{r^2\sin^2\theta}\frac{\partial v_\phi}{\partial \phi}\right), \tag{1.26}$$

$$\frac{\partial v_\phi}{\partial t} + v_r\frac{\partial v_\phi}{\partial r} + \frac{v_\theta}{r}\frac{\partial v_\phi}{\partial \theta} + \frac{v_\phi}{r\sin\theta}\frac{\partial v_\phi}{\partial \phi} + \frac{v_\phi v_r}{r} + \frac{v_\theta v_\phi\cot\theta}{r}$$
$$= -\frac{1}{\rho}\frac{1}{r\sin\theta}\frac{\partial p}{\partial \phi} + F_\phi + \nu\left(\nabla^2 v_\phi - \frac{v_\phi}{r^2\sin^2\theta} + \frac{2}{r^2\sin\theta}\frac{\partial v_r}{\partial \phi} + \frac{2\cos\theta}{r^2\sin^2\theta}\frac{\partial v_\theta}{\partial \phi}\right), \tag{1.27}$$

with the continuity equation

$$\frac{1}{r^2}\frac{\partial}{\partial r}(r^2 v_r) + \frac{1}{r\sin\theta}\frac{\partial}{\partial\theta}(v_\theta\sin\theta) + \frac{1}{r\sin\theta}\frac{\partial v_\phi}{\partial\phi} = 0. \qquad (1.28)$$

The components of vorticity are given by

$$\omega_r = \frac{1}{r\sin\theta}\left\{\frac{\partial}{\partial\theta}(v_\phi\sin\theta) - \frac{\partial v_\theta}{\partial\phi}\right\},$$

$$\omega_\theta = \frac{1}{r\sin\theta}\frac{\partial v_r}{\partial\phi} - \frac{1}{r}\frac{\partial}{\partial r}(r v_\phi), \qquad (1.29)$$

$$\omega_\phi = \frac{1}{r}\frac{\partial}{\partial r}(r v_\theta) - \frac{1}{r}\frac{\partial v_r}{\partial\theta}.$$

The Stokes stream function, for a rotationally symmetric flow independent of ϕ, is now defined such that

$$v_r = -\frac{1}{r^2\sin\theta}\frac{\partial\psi}{\partial\theta}, \quad v_\theta = \frac{1}{r\sin\theta}\frac{\partial\psi}{\partial r}. \qquad (1.30)$$

2

Steady flows bounded by plane boundaries

2.1 Plane Couette–Poiseuille flow

We begin this discussion with a consideration of the simplest of flows, namely that between parallel plane boundaries under the action of a pressure gradient parallel to the boundaries in, say, the x-direction with one boundary sliding in that direction. It suits our purpose to generalise the situation by allowing fluid to be injected with uniform constant velocity at one boundary, and similarly removed at the other. Such flows, driven solely by a pressure gradient, are associated with Poiseuille (1840) who was concerned with the flow in tubes as discussed in the next chapter. Flows driven by the sliding or 'scraping' motion of the boundary were considered by Couette (1890).

With x measured in the flow direction, and y perpendicular to the plane boundaries, assumed to be at $y = 0, h$, we have $\mathbf{v} = (u, v, 0)$ and we assume that $u = u(y)$, $v = v(y)$. That is, we assume the flow to be fully developed, and so independent of any entry conditions to the channel. The Navier–Stokes equations, then, are to be solved subject to the conditions

$$u = 0, \quad v = V \quad \text{at} \quad y = 0,$$
$$u = U, \quad v = V \quad \text{at} \quad y = h. \tag{2.1}$$

From (1.12) we see at once that $\partial v/\partial y = 0$, and hence $v \equiv V$. In the absence of any body force equation (1.10) shows that $\partial p/\partial y = 0$ so that $p = p(x)$, and the equation we must solve, namely (1.9), reduces to

$$V\frac{\partial u}{\partial y} = -\frac{1}{\rho}\frac{\partial p}{\partial x} + v\frac{\partial^2 u}{\partial y^2}. \tag{2.2}$$

From the assumptions we have made, it is clear from (2.2) that the pressure gradient is a constant, and we set $\partial p/\partial x = -P$. The solution, then, of (2.2) subject to (2.1) is

$$u(y) = \frac{P}{\rho V} y + U \left(1 - \frac{Ph}{\rho U V}\right) \left(1 - e^{Vy/\nu}\right) / \left(1 - e^{R}\right), \qquad (2.3)$$

where $R = Vh/\nu$ is a Reynolds number. In the limit as $V \to 0$, when the boundaries are impermeable, we recover the classical Couette–Poiseuille solution

$$u(y) = U\frac{y}{h} + \frac{1}{2}\frac{Ph^2}{\mu}\left\{\frac{y}{h} - \left(\frac{y}{h}\right)^2\right\}. \qquad (2.4)$$

If we define Q as the mass flux along the channel per unit depth, then we have

$$Q = \int_0^h \rho u \, \mathrm{d}y = \frac{1}{2}\rho U h\left(1 + \frac{Ph^2}{6\mu U}\right). \qquad (2.5)$$

For $U < 0$ there is a region close to $y = h$ in which the flow is reversed in direction, whilst for U sufficiently large and negative, such that $U < -Ph^2/6\mu$, there is a net mass flux in the negative x-direction.

We note that if $U = Ph^2/2\mu$ then $\partial u/\partial y = 0$ at $y = h$, which implies that the boundary $y = h$ behaves as a free surface. Indeed, if we let $P = \rho g \sin\alpha$, where g is the acceleration due to gravity, we have, in that case,

$$u(y) = \frac{\rho g h^2 \sin\alpha}{2\mu}\left\{2\frac{y}{h} - \left(\frac{y}{h}\right)^2\right\}, \qquad (2.6)$$

which represents the fully developed flow of a film of viscous fluid, thickness h, down a plane inclined at angle α to the horizontal in the presence of the body force per unit mass $\mathbf{F} = (-g\sin\alpha, -g\cos\alpha, 0)$. The free-surface speed is $\rho g h^2 \sin\alpha/2\mu$, and $Q = \rho g h^3 \sin\alpha/3\nu$. Sample velocity profiles are shown in figure 2.1.

Returning next to the case of permeable walls, for which $v \equiv V$ and u is as in equation (2.3), we have for the limiting situation $V \gg \nu/h$, $Ph/\rho U$, the velocity $u(y) \approx U\exp\{V(y - h)/\nu\}$ so that u is essentially zero except close to $y = h$, specifically where $h - y = O(\nu/V)$ where u increases rapidly to the boundary velocity U. Conversely, if we reverse the injection velocity, and write $V = -\overline{V}$, then for $\overline{V} \gg \nu/h$, $Ph/\rho U$ we have $u(y) \approx U\{1-\exp(-\overline{V}y/\nu)\} \approx U$ except for $y = O(\nu/\overline{V})$ where again a rapid transition takes place, in a thin 'boundary layer', to enable satisfaction of the no-slip condition. In this latter case fluid particles emerge into the fluid, at $y = h$, with momentum $U\mathbf{i} - \overline{V}\mathbf{j}$ per unit mass. This remains unchanged, by either pressure or viscous forces, until viscous stresses destroy the x-component close to $y = 0$. The boundary vorticity $\zeta_w = -U\overline{V}/\nu$, and as we have seen this penetrates, by diffusion, a distance only $O(\nu/\overline{V})$ from the boundary, restrained as it is by convection towards it.

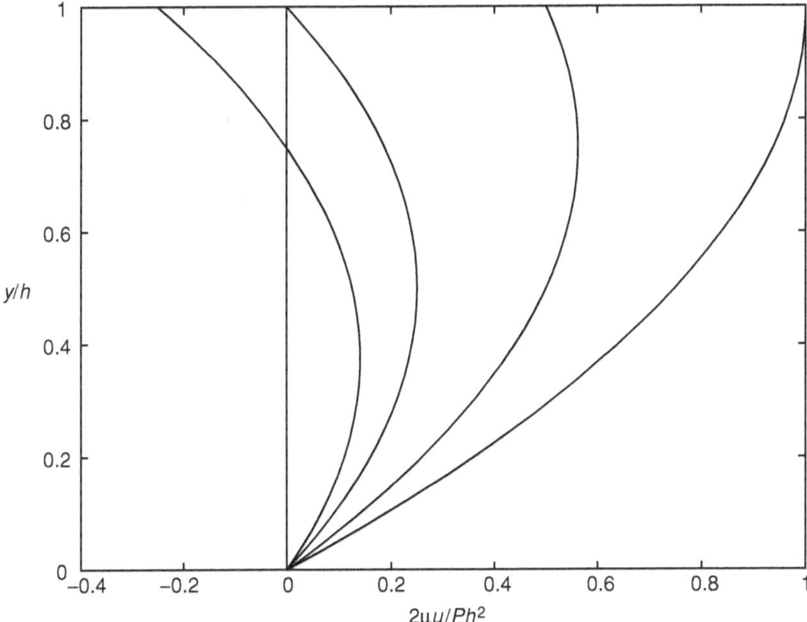

Figure 2.1 Velocity profiles in Couette–Poiseuille flow from equation (2.4) with, from the left, $2\mu U/Ph^2 = -1/4, 0, 1/2, 1$ respectively.

Kolmogorov (see Meshalkin and Sinai 1961) discussed a special class of forced flows in his seminar in Moscow on selected problems in analysis. These are the class of flows for which $\mathbf{F} = \gamma \sin(\pi y/h)\mathbf{i}$, $-h \leq y \leq h$ so that $p = \text{constant}$, $u = (\gamma h^2/\pi^2 \nu)\sin(\pi y/h)$ and $Q = 0$. Whilst the forcing and flow are artificial, the flow has served as a prototype to model spatially periodic flows, such as waves, and their instability.

Several solutions for flow in channels of finite cross-section, S, bounded by plane boundaries, for which $\mathbf{v} = \{u(y, z), 0, 0\}$, are available. For such situations we have, from equations (1.9) to (1.12) and in the absence of body forces

$$\frac{\partial^2 u}{\partial y^2} + \frac{\partial^2 u}{\partial z^2} = -\frac{P}{\mu}. \tag{2.7}$$

Consider first the rectangular channel $0 \leq y \leq h$, $0 \leq z \leq l$. If we write

$$u(y, z) = \frac{1}{2}\frac{Ph^2}{\mu}\left\{\frac{y}{h} - \left(\frac{y}{h}\right)^2\right\} + f(y, z),$$

that is represent the solution as a perturbation of plane Poiseuille flow, then $f(y, z)$ is a harmonic function for which $f = 0$ on $y = 0, h$ and $f = -(Ph^2/2\mu)\{(y/h) - (y/h)^2\}$ on $z = 0, l$. From Boussinesq (1868) we have the solution as

$$u(y, z) = \frac{1}{2}\frac{Ph^2}{\mu}\left\{\frac{y}{h} - \left(\frac{y}{h}\right)^2\right\}$$

$$-\frac{4Ph^2}{\mu\pi^3}\sum_{n=0}^{\infty}\frac{\left[\sinh\left\{\frac{(2n+1)\pi z}{h}\right\} + \sinh\left\{\frac{(2n+1)\pi(l-z)}{h}\right\}\right]\sin\left\{\frac{(2n+1)\pi y}{h}\right\}}{(2n+1)^3\sinh\left\{\frac{(2n+1)\pi l}{h}\right\}}, \qquad (2.8)$$

which gives the mass flux along the channel as

$$Q = \int\!\!\int_S \rho u(y, z)\, dy\, dz = \frac{Ph^3 l}{12\nu} - \frac{16Ph^4}{\pi^5\nu}\sum_{n=0}^{\infty}\frac{\left[\cosh\left\{\frac{(2n+1)\pi l}{h}\right\} - 1\right]}{(2n+1)^5\sinh\left\{\frac{(2n+1)\pi l}{h}\right\}}.$$

$$(2.9)$$

From (2.8) and (2.9) the results for plane Poiseuille flow may be recovered in the limit $l \to \infty$. Rowell and Finlayson (1928) also considered the channel of rectangular cross-section with one of the bounding walls moving in the flow direction. This may be considered as a generalisation of the Couette–Poiseuille flow (2.4).

Triangular cross-sections have also received attention. Boussinesq (1868) considered polynomial solutions of the form

$$u(y, z) = -\frac{P}{4\mu h}(y - h)(y - az)(y - bz),$$

which vanishes on the three sides of a triangle. Direct substitution in (2.7) shows at once that $a = -b = \sqrt{3}$ is the only possibility; thus the triangle is an equilateral triangle of side $2h/\sqrt{3}$, with

$$u = -\frac{P}{4\mu h}(y - h)(y^2 - 3z^2).$$

The corresponding mass flux is given by

$$Q = \frac{Ph^4}{60\sqrt{3}\nu}. \qquad (2.10)$$

Proudman (1914) considered the flow in the right-angled isosceles triangle $y = \pi, y \pm z = 0$, to find

$$u(y, z) = \frac{P}{2\mu}(y + z)(\pi - y) - \frac{P}{\pi\mu}\sum_{n=0}^{\infty}\frac{1}{(n + \frac{1}{2})^3 \sinh\{(2n + 1)\pi\}}$$

$$\times [\sinh\{(n + \tfrac{1}{2})(2\pi - y + z)\}\sin\{(n + \tfrac{1}{2})(y + z)\} - \sinh\{(n + \tfrac{1}{2})(y + z)\}$$

$$\times \sin\{(n + \tfrac{1}{2})(y - z)\}], \qquad (2.11)$$

so that

$$Q = \frac{P}{2v}\left\{\frac{\pi^4}{6} - \frac{1}{\pi}\sum_{n=0}^{\infty}[\coth\{(2n+1)\pi\} + \operatorname{cosec}\{(2n+1)\pi\}](n+\tfrac{1}{2})^{-5}\right\}.$$

(2.12)

2.2 Beltrami flows and their generalisation

Flows associated with the name of Beltrami (1889) (though Truesdell (1954, section 52) notes that nearly all the work of Beltrami had been anticipated by Gromeka (1881)) are those for which $\mathbf{v} \wedge \boldsymbol{\omega} = \mathbf{0}$ so that $\boldsymbol{\omega} = \alpha\mathbf{v}$ for some scalar function α. In that case the vorticity equation (1.7) reduces to

$$\frac{\partial \boldsymbol{\omega}}{\partial t} = \nabla \wedge \mathbf{F} + v\nabla^2\boldsymbol{\omega},$$

(2.13)

so that, as noted by Trkal (1919), the only steady Beltrami flows of a viscous fluid, apart from those that are irrotational, are those sustained by a non-conservative body force. Otherwise the vorticity satisfies the diffusion equation and decays. We shall consider unsteady Beltrami flows in chapter 4. A generalisation of Beltrami flows requires, not $\mathbf{v} \wedge \boldsymbol{\omega} = \mathbf{0}$ but

$$\nabla \wedge (\mathbf{v} \wedge \boldsymbol{\omega}) = \mathbf{0}$$

(2.14)

which for steady flow, in the absence of any non-conservative body force, also requires from (1.7)

$$\nabla^2\boldsymbol{\omega} = \nabla \wedge \nabla \wedge \boldsymbol{\omega} = \mathbf{0}.$$

(2.15)

If we introduce the stream function ψ, as in equation (1.14), such that $\mathbf{v} = \nabla \wedge (\psi\mathbf{k})$ then, with $\boldsymbol{\omega} = (0, 0, \zeta)$ we have, from equation (2.14),

$$\frac{\partial}{\partial x}\left(\zeta\frac{\partial \psi}{\partial y}\right) - \frac{\partial}{\partial y}\left(\zeta\frac{\partial \psi}{\partial x}\right) = 0.$$

(2.16)

Further, as noted in (1.15), we have

$$\frac{\partial^2 \psi}{\partial x^2} + \frac{\partial^2 \psi}{\partial y^2} = -\zeta,$$

(2.17)

and equation (2.15) requires

$$\frac{\partial^2 \zeta}{\partial x^2} + \frac{\partial^2 \zeta}{\partial y^2} = 0.$$

(2.18)

The solution of equation (2.16) is

$$\zeta = -f(\psi).$$

(2.19)

An exact solution of the Navier–Stokes equations results when the three equations (2.17) to (2.19) are compatible. A special case is when $f(\psi) = K$, a constant, and the equations then reduce to

$$\frac{\partial^2 \psi}{\partial x^2} + \frac{\partial^2 \psi}{\partial y^2} = K, \tag{2.20}$$

which corresponds to a flow in which vorticity is uniform. Such a flow, with uniform vorticity, is a shear flow, with a linearly varying velocity $u(y)$, say, to which we may add a potential flow. Tsien (1943), in his study of the shear flow past aerofoils, gave the solution for a line source and a line vortex, respectively, embedded in such a shear flow as

$$\psi = ay + by^2 + c \tan^{-1}\left(\frac{y}{x}\right), \tag{2.21}$$

$$\psi = ay + by^2 + c \ln(x^2 + y^2), \tag{2.22}$$

for constants a, b, c, whilst Wang (1990b) presented the solution

$$\psi = ay^2 + b\,e^{-\lambda y} \cos \lambda x, \tag{2.23}$$

with λ constant, which he interpreted in terms of a shear flow over convection cells.

A further solution given by Wang (1991) is

$$\psi = y(ay + bx), \tag{2.24}$$

which may be interpreted as the oblique impingement of two flows for which $y = 0$, $y = -bx/a$ are the dividing streamlines.

The solutions (2.21) to (2.24) are, of course, trivial in the sense that they do not involve the viscosity. Wang (1991) showed that the above range of solutions may be generalised by writing

$$\zeta = \psi + B(x, y),$$

and assuming that B is linear so that $B(x, y) = ax + by$ (in fact Wang has $a \equiv 0$); the vorticity equation (1.17) then yields, as an equation for ψ,

$$\frac{\partial^2 \psi}{\partial x^2} + \frac{b}{\nu}\frac{\partial \psi}{\partial x} = -\left(\frac{\partial^2 \psi}{\partial y^2} - \frac{a}{\nu}\frac{\partial \psi}{\partial y}\right). \tag{2.25}$$

As we see, this equation retains both the viscous diffusion terms and the linearised convection terms; in that sense equation (2.20) is a special case. Solutions of the linear equation (2.25), which embrace a wide selection of recorded exact solutions of the Navier–Stokes equations, are available by the method of separation of variables. Thus, if we set $\psi(x, y) = F(x)G(y)$ then F, G

respectively, satisfy

$$\frac{d^2 F}{dx^2} + \frac{b}{v}\frac{dF}{dx} - CF = 0, \qquad \frac{d^2 G}{dy^2} - \frac{a}{v}\frac{dG}{dy} + CG = 0, \qquad (2.26)$$

where C is the separation constant. We consider now the consequences of (2.26) in several special cases.

2.2.1 Flow downstream of a grid

If we take $a = 0$, $b = -U$ and a positive separation constant $C = \lambda^2$ then we have

$$G = A \sin \lambda y \quad \text{and} \quad F = \exp\left[\left\{U/v\lambda \pm \sqrt{(U^2/v^2\lambda^2 + 4)}\right\}\lambda x/2\right].$$

Choosing, now, the solution that decays in the x-direction, and setting $\lambda = 2\pi/h$ gives

$$\psi = Uy + A \sin(2\pi y/h) \exp\left[\left\{R - \sqrt{(R^2 + 16\pi^2)}\right\}x/2h\right], \qquad (2.27)$$

where $R = Uh/v$ is the Reynolds number. Kovásznay (1948) obtained this solution which he suggested may be used to describe the flow downstream from a two-dimensional grid with grid spacing h. With $R = 40$ the streamlines for one period are shown in figure 2.2.

In an extension of this Lin and Tobak (1986) take $C = -\lambda^2 < 0$. If we again set $\lambda = 2\pi/h$, then Lin and Tobak's solution may be written as

$$\psi = Uy + A \exp\left[-2\pi y/h + \left\{R + \sqrt{(R^2 - 16\pi^2)}\right\}x/2h\right]. \qquad (2.28)$$

Lin and Tobak interpret this solution as the flow over the permeable plate $y = 0$, $x < 0$ at which suction is applied, noting that for sufficiently large values of A/U there will be flow reversal. However, it should also be noted that the solution (2.28) does not satisfy the no-slip condition at the plate.

A slightly more general solution than (2.28) has

$$\psi = Uy - A \sinh(2\pi y/h) \exp\left[\left\{R + \sqrt{(R^2 - 16\pi^2)}\right\}x/2h\right],$$

which Wang (1966) interprets as a uniform stream encountering a rotational, diverging counterflow.

2.2.2 Flow due to a stretching plate

We now take $a = -U$, $b = 0$ and the separation constant $C = 0$ so that, in general, the solution for ψ is

$$\psi = Ux + (-Ux + k_1)(k_2 + k_3\, e^{-Uy/v}).$$

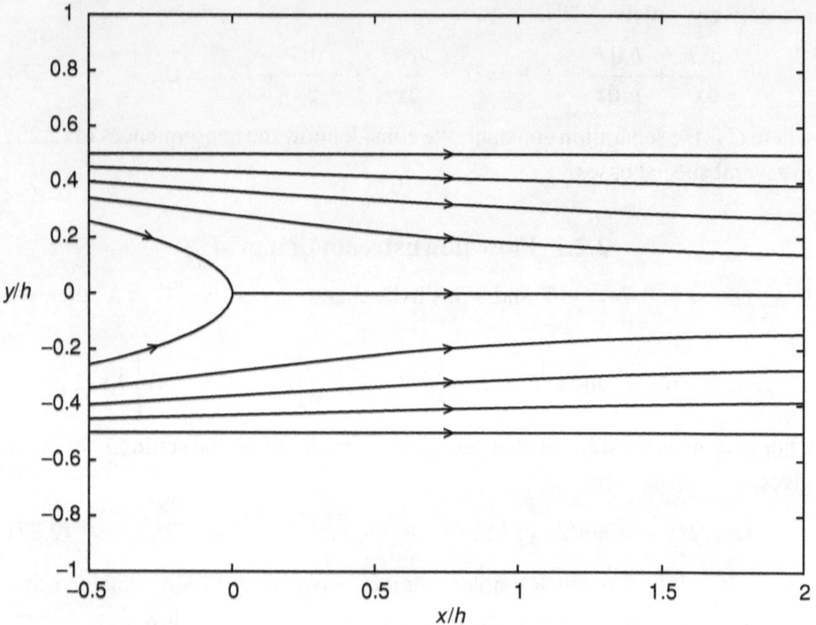

Figure 2.2 Streamlines of the flow represented by equation (2.27) with $A = -(2\pi)^{-1}$ and $R = 40$.

Applying the boundary conditions $\partial\psi/\partial x = 0$ on $y = 0$, $\partial\psi/\partial y = 0$ on $x = 0$ and $\partial\psi/\partial x \to -U$ as $y \to \infty$ gives

$$\psi = Ux(1 - e^{-Uy/\nu}). \tag{2.29}$$

With

$$u = (U^2 x/\nu)\, e^{-Uy/\nu} \quad \text{and} \quad v = U(e^{-Uy/\nu} - 1),$$

we have a flow in the half-plane $y \geq 0$ with fluid speed $U^2 x/\nu$ along the boundary $y = 0$. The solution (2.29) was discovered by Riabouchinsky in 1924, noted in passing as an exact solution in a quite different context by Stuart (1966a) and interpreted by Crane (1970) as the flow due to the stretching plate $y = 0$. Note how vorticity created at the stretching plate is confined to a region of thickness $O(\nu/U)$, beyond which we have uniform flow towards it. Danberg and Fansler (1976) have introduced a uniform flow parallel to the plate, whilst Wang (1984) has extended the analysis to the case in which the plate stretches in orthogonal directions.

McLeod and Rajagopal (1987) have established the uniqueness of this solution, a practical application of which we shall encounter in the next section.

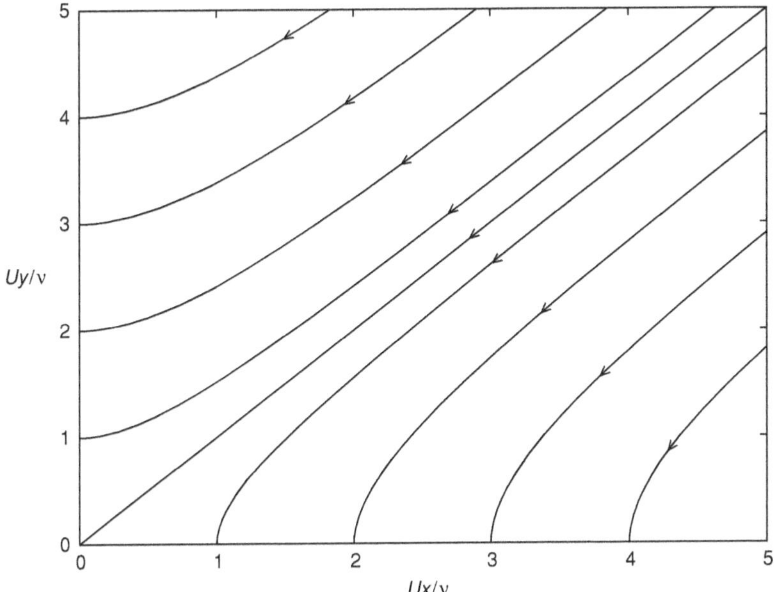

Figure 2.3 The corner flow streamlines represented by equation (2.30).

2.2.3 Flow into a corner

With $a = -U$, $b = U$ and $C = 0$ we have, as the solution which satisfies the no-slip condition on $x = y = 0$, for the flow in the quarter-plane $x, y \geq 0$, bounded as $x, y \to \infty$,

$$\psi = U(x - y) + \nu(e^{-Ux/\nu} - e^{-Uy/\nu}). \qquad (2.30)$$

At large distances from the bounding walls the streamlines are given by the parallel lines $y = x + \text{constant}$. With

$$u = -U(1 - e^{-Uy/\nu}) \quad \text{and} \quad v = -U(1 - e^{-Ux/\nu}),$$

we see that the no-slip condition is satisfied at each of the boundaries, which are permeable, and at each of which there is suction. The streamline pattern is shown in figure 2.3. The solution (2.30) is a special case of a more general class of solutions considered by Berker (1963, section 15).

2.2.4 The asymptotic suction profile

A solution closely related to (2.30) is obtained by taking $a = -V$, $b = -U$, $C = 0$. The solution of (2.25), in the half-plane $y \geq 0$ and bounded at infinity, which satisfies the no-slip condition on $y = 0$ is

$$\psi = Vx + Uy + \frac{Uv}{V}e^{-Vy/v}, \tag{2.31}$$

for which $u = U(1 - e^{-Vy/v})$, $v \equiv -V$. The solution (2.31) corresponds, therefore, to uniform flow over an infinite porous plate at which the suction velocity has magnitude V.

This solution may be interpreted as the steady flow far downstream from the leading edge of a semi-infinite porous flat plate. In the absence of suction, the vorticity created at the surface would diffuse indefinitely as we go downstream. The presence of suction inhibits this and the balance between convection and diffusion, represented by the solution (2.31), confines the vorticity, as in the two previous examples, to a layer of thickness $O(v/U)$.

2.3 Stagnation-point flows

2.3.1 The classical Hiemenz (1911) solution

When a steady stream of a viscous fluid approaches a rigid stationary cylinder, the stream is brought to rest at the surface of the body and divides about it. Although the fluid is at rest, at each point of the surface of the cylinder, by analogy with the flow of an inviscid fluid, we identify stagnation points as those points on the surface at which the stream attaches to, or separates from, the cylinder. The flow in the neighbourhood of a stagnation point of attachment may be modelled by the flow towards an infinite rigid flat plate. Now, for an inviscid fluid, the irrotational flow against the flat plate $y = 0$ is well known to be $u = kx$, $v = -ky$. The constant k is not directly relevant to the flow pattern close to the stagnation point, and is proportional to the free-stream speed about the cylinder. The inviscid stream function is $\psi = kxy$. In his study of the flow of a viscous fluid at a stagnation point it would appear to have been natural for Hiemenz (1911) to have assumed $\psi(x, y) = xF(y)$. If we introduce dimensionless variables, noting that there is no natural length scale in this problem, then

$$\psi = (vk)^{1/2}xf(\eta) \quad \text{where} \quad \eta = \left(\frac{k}{v}\right)^{1/2} y, \tag{2.32}$$

and from (1.17) we have, as the equation for f,

$$f^{iv} + ff''' - f'f'' = 0, \tag{2.33}$$

where a prime denotes differentiation with respect to η, together with the boundary conditions

$$f(0) = f'(0) = 0, \quad f'(\infty) = 1. \tag{2.34}$$

Integrating (2.33) once gives

$$f''' + ff'' - f'^2 + 1 = 0, \tag{2.35}$$

and from (1.9), (1.10) we may infer that the pressure

$$\frac{p_0 - p}{\rho} = \frac{1}{2}k^2 x^2 + \frac{1}{2}\nu k f^2 + \nu k f', \tag{2.36}$$

where p_0 is a constant. Unlike the previous solutions of section 2.2 the present solution describes a flow in which linear diffusion of vorticity is balanced by *non-linear* convection of vorticity. The solution of (2.35), subject to the conditions (2.34), has been calculated numerically by Hiemenz (1911) and by Howarth (1934).

Rott (1956) extended the solution of Hiemenz to include the situation in which the plane boundary slides in its own plane in the x-direction. This provides, for example, a model of the flow in the neighbourhood of the stagnation point of a rotating circular cylinder placed in a uniform stream. In place of (2.32) we have

$$\psi = (\nu k)^{1/2} x f(\eta) + u_w \left(\frac{\nu}{k}\right)^{1/2} \int_0^\eta g \, d\eta, \tag{2.37}$$

where u_w is the speed of the translating plate. With p unchanged as in (2.36), equation (1.9) then yields, as equation for g,

$$g'' + fg' - f'g = 0 \quad \text{with} \quad g(0) = 1, \quad g(\infty) = 0. \tag{2.38}$$

Rott identifies the solution of (2.38) as $g = f''(\eta)/f''(0)$. In other words the superimposed cross flow is proportional to the shear distribution of the Hiemenz flow. In figure 2.4 we show both $f'(\eta)$ and $g(\eta)$.

An investigation by Becker (1976) provides a novel application of the Hiemenz type of flow. Becker considers the situation in which a film of liquid, of constant thickness h, flows away from the stagnation point on a cylindrical surface with constant radius of curvature R_c, and a horizontal axis. The film is sustained by rain falling with constant speed V; the 'rain' is modelled as a homogeneous medium with density $\epsilon\rho$, where ρ is the liquid density and $0 < \epsilon < 1$. If α is the angle of inclination of the tangent of the cylinder surface to the horizontal, and x is measured from the stagnation point along the surface, then $\alpha = x/R_c$ and Becker makes the assumption $\alpha \ll 1$.

In this problem we have a body force per unit mass $Y\mathbf{j} = -g\mathbf{j}$, so that in (1.9), (1.10) it is convenient to work with the modified pressure $p^* = p + \rho g y$.

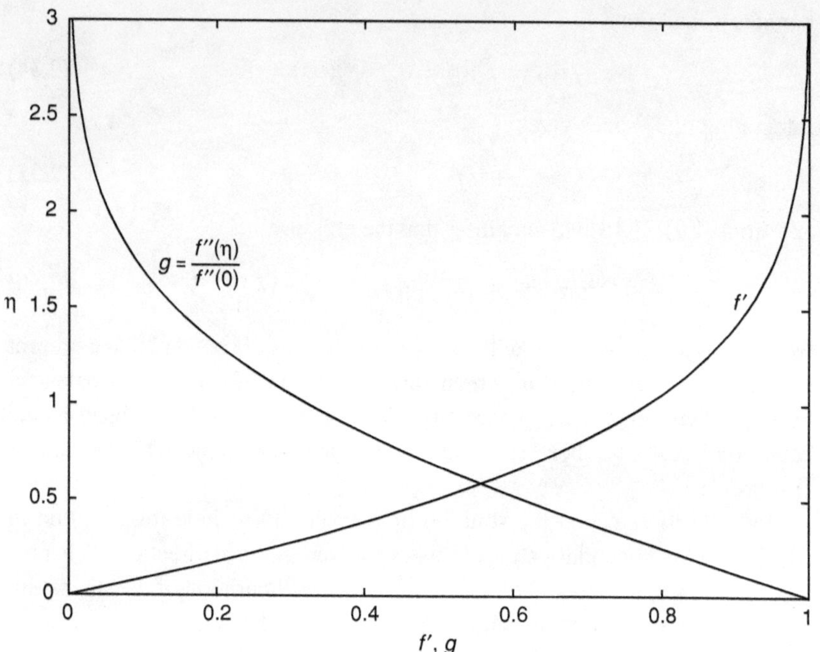

Figure 2.4 The stagnation-point velocity profile $f'(\eta)$, and the cross-flow profile $g(\eta) = f''(\eta)/f''(0)$.

The same similarity transformation (2.32) as used in the classical Hiemenz flow is adopted but now, in addition to the standard conditions at the boundary, we must consider conditions at the interface $y = h$. If we define $\eta_h = (k/\nu)^{1/2}h$ then mass conservation requires

$$f(\eta_h) = \epsilon V/(\nu k)^{1/2}. \tag{2.39}$$

The balance of tangential momentum at the interface, in the formal limit $R_c \to \infty$, requires

$$f''(\eta_h) + f(\eta_h)f'(\eta_h) = 0. \tag{2.40}$$

For the pressure note that as before we have

$$\frac{p^*}{\rho} = \frac{1}{2}bx^2 - \frac{1}{2}\nu k f^2 - \nu k f', \tag{2.41}$$

where we have set $p_0 = 0$, and b is, as yet, undetermined. A consideration of the balance of momentum normal to the interface leads to the result that $b = -g/R_c$, where higher powers of R_c^{-1} have been ignored. We now define a Reynolds number $Re = (gR_c)^{1/2}R_c/\nu$ and a Froude number $Fr = \epsilon V/(gR_c)^{1/2}$.

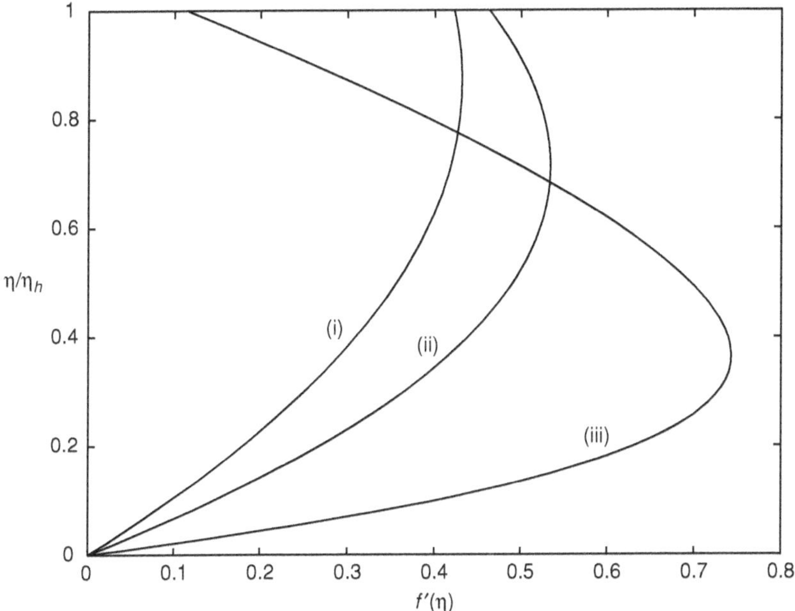

Figure 2.5 Solutions of equation (2.42) corresponding to (i) $\lambda = 1.2$, (ii) $\lambda = 1.0$, (iii) $\lambda = 0.8$.

If in the limit $R_c \to \infty$, $Fr\,Re^{1/2} \to \beta$, a constant, then we have an exact solution of the Navier–Stokes equations where $f(\eta)$ satisfies

$$f''' + ff'' - f'^2 + \lambda = 0, \tag{2.42}$$

where λ is determined from β. This equation is to be solved subject to $f(0) = f'(0) = 0$, and the interface condition (2.40). Some typical velocity profiles are shown in figure 2.5.

Becker's analysis included finite curvature effects, as the model of a road, for which the above may be considered as the leading term in an appropriate expansion.

2.3.2 Oblique stagnation-point flows

In section 2.3.1 above, the dividing, or stagnation, streamline intersects the plane boundary $y = 0$ orthogonally. There is, however, a class of stagnation-point flows for which the dividing streamline intersects the boundary at an arbitrary angle. Consider first an inviscid fluid and the stream function

$$\psi = kxy + \tfrac{1}{2}\varsigma_0 y^2, \tag{2.43}$$

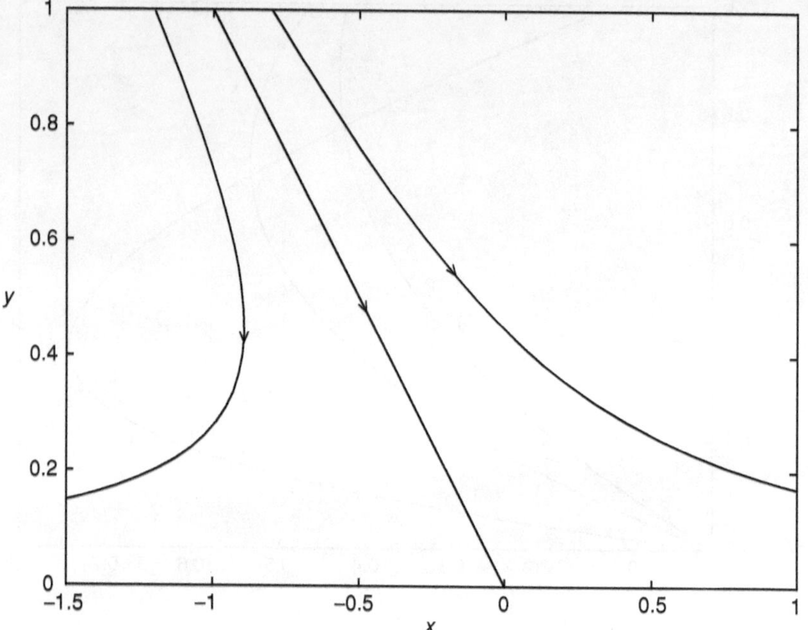

Figure 2.6 Streamlines for oblique inviscid stagnation-point flow, from equation (2.43) with $\zeta_0 = 2k$.

which combines both the classical stagnation-point flow and a cross flow of uniform shear, that is of constant vorticity $-\zeta_0$. This stream function satisfies both the Euler equations and the Navier–Stokes equations, though not the viscous condition of no slip at $y = 0$. Taking $y = 0$ as the boundary again, we see from (2.43) that the dividing streamline is now $y = -2(k/\zeta_0)x$ and the flow is sketched in figure 2.6.

This form of stagnation-point flow has been developed by Stuart (1959), Tamada (1979) and Dorrepaal (1986) for a viscous fluid as follows. With a superposed cross flow present it is natural, taking account of (2.37) and (2.43), to write the stream function as

$$\psi = (\nu k)^{1/2} x f(\eta) + \zeta_0 \left(\frac{\nu}{k}\right) \int_0^\eta g \, d\eta, \tag{2.44}$$

where again $\eta = (k/\nu)^{1/2} y$. Substitution of (2.44) into equation (1.17) yields ordinary differential equations for f and g. Integrating each of these once with respect to η results in equation (2.35) for f, whilst for g we have

$$g'' + fg' - f'g = \text{constant}, \tag{2.45}$$

together with $g(0) = 0$, $g'(\infty) = 1$. Now, the solution for f has the property that, as $\eta \to \infty$, $f \sim \eta - A$, where $A = 0.6479$ represents the viscous

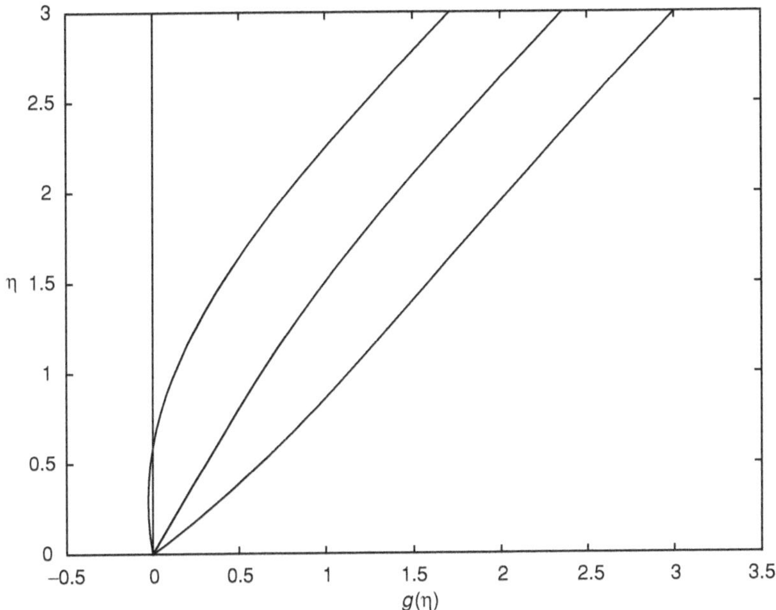

Figure 2.7 Cross-flow velocity profiles $g(\eta)$ for oblique stagnation-point flow with, from the left, $B - A = 0.6479, 0, -0.6479$, respectively.

displacement effect. Similarly we have $g \sim \eta - B$, where B is undetermined, so that (2.45) becomes

$$g'' + fg' - f'g = B - A. \tag{2.46}$$

From equations (1.9) and (1.10) the pressure is given by

$$\frac{p_0 - p}{\rho} = \frac{1}{2}k^2x^2 + \frac{1}{2}\nu k f^2 + \nu k f' - \zeta_0(\nu k)^{1/2}(B - A)x. \tag{2.47}$$

With B arbitrary we see that (2.47) incorporates an additional constant pressure gradient in the x-direction, proportional to $B - A$, the effect of which is, simply, to determine the displacement of the uniform shear flow parallel to the plate. Denoting $g'(0) = D$, where D is known when B is specified, $f''(0) = C = 1.2326$ the solution of equation (2.46) is formally obtained as

$$g(\eta) = (A - B)f'(\eta) + C\{D - C(A - B)\}f''(\eta)\int_0^\eta \{f''(t)\}^{-2}$$

$$\times \exp\left\{-\int_0^t f(s)\,ds\right\}dt. \tag{2.48}$$

In figure 2.7 we show, for various values of $B - A$, profiles $g(\eta)$. In their analyses Stuart and Tamada take $B = A$, and Dorrepaal takes $B = 0$.

From a consideration of the solution (2.44) close to the boundary $y = 0$, it can be shown that the dividing streamline $\psi = 0$ meets the boundary at

$$x = x_s = -\zeta_0 \left(\frac{\nu}{k^3}\right)^{1/2} \left(\frac{D}{C}\right),$$

where its slope is given by

$$-\frac{3kC^2}{\zeta_0\{(B - A)C + D\}},$$

which is independent of the kinematic viscosity ν. Furthermore, the ratio of this slope to that of the dividing streamline far from the boundary, where viscous effects are unimportant, is

$$\frac{3}{2} \frac{C^2}{\{(B - A)C + D\}}$$

which is, in addition, independent of k and ζ_0, depending only upon the constant pressure gradient parallel to the boundary through both B and D.

2.3.3 Two-fluid stagnation-point flow

The classical Hiemenz stagnation-point flow is that of a stream against a solid boundary. Wang (1985a) has extended this to the flow against the interface with a second fluid. The interface is assumed planar so that we may expect the surface tension to be large, or the density of the lower fluid to be much greater than the density of the upper fluid.

The similarity of the Hiemenz flow is preserved so that in the upper fluid the stream function is expressed as in (2.32), with $f(\eta)$ again satisfying equation (2.35). However the no-slip condition is violated in this case and we have $f'(0) = \lambda$, where λ is determined only following a consideration of the flow in the second fluid in $y < 0$. In the second fluid, which we assume has density $\bar{\rho}$ and viscosity $\bar{\mu}$, we write

$$\bar{\psi} = -(\bar{\nu}k\lambda)^{1/2}x\,\bar{f}(\bar{\eta}) \quad \text{where now} \quad \bar{\eta} = -\left(\frac{k\lambda}{\bar{\nu}}\right)^{1/2} y. \tag{2.49}$$

Since we require $\bar{f}'(\infty) = 0$, where a prime again denotes differentiation with respect to the independent variable, the equation for $\bar{f}$ is

$$\bar{f}''' + \bar{f}\bar{f}'' - \bar{f}'^2 = 0, \quad \bar{f}(0) = 0, \quad \bar{f}'(0) = 1, \quad \bar{f}'(\infty) = 0. \tag{2.50}$$

Where the conditions at $\bar{\eta} = 0$ ensure continuity of velocity at the interface. The solution of (2.50) is

$$\bar{f}(\bar{\eta}) = 1 - e^{-\bar{\eta}}.$$

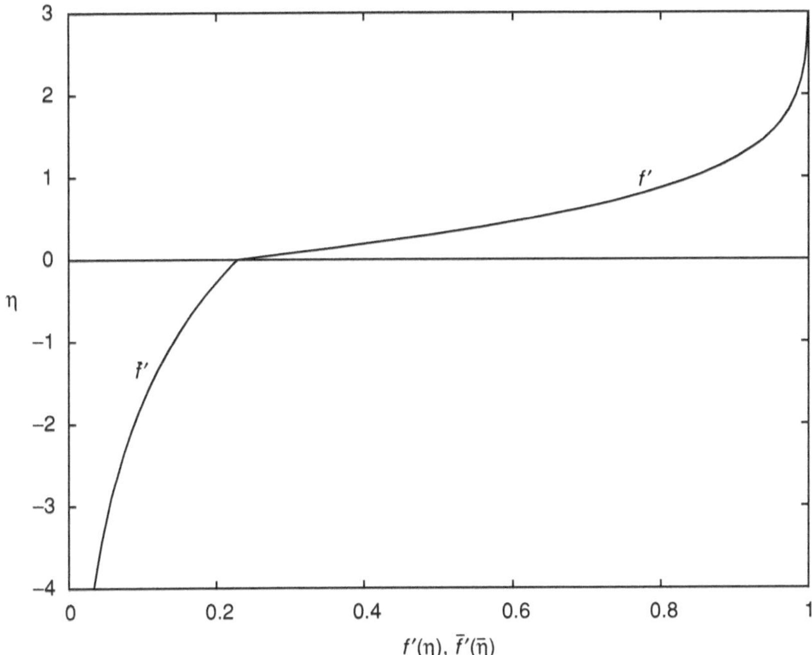

Figure 2.8 Tangential velocity profile for a two-fluid stagnation-point flow.

This solution we have met before in equation (2.29). There, as remarked, Crane (1970) interpreted it as the flow due to a stretching plate. Whilst this might be difficult to envisage physically, in the present context the 'stretching plate' is simply the free surface at which the velocity does increase linearly with distance. To complete the solution the parameter λ must be determined, and this is possible from continuity of stress at the interface from which we have

$$\frac{f''(0)}{\lambda^{3/2}} = \frac{\bar{\rho}}{\rho}\left(\frac{\bar{\nu}}{\nu}\right)^{1/2}, \tag{2.51}$$

so that λ may be determined from the numerical solution of (2.35), (2.51). For a two-fluid system with water over olive oil we have, in cgs units, $\rho = \bar{\rho} = 1$, $\nu = 0.0114$, $\bar{\nu} = 1.0$, and find $\lambda = 0.22841$. The tangential velocity profile is shown in figure 2.8.

Wang (1987) has also considered the impingement of two stagnation-point flows at a two-fluid interface, whilst Tilley and Weidman (1998) have extended Wang's analysis to oblique stagnation-point flow in a two-fluid system.

2.4 Channel flows

At the beginning of this chapter we considered the flow in parallel-sided channels, typical of which is the classical Couette–Poiseuille flow. With a cross flow in the channel due to equal suction and injection at the boundaries, convective terms were retained, but only linearly. We now consider channel flows in which the non-linear terms can be neither neglected nor linearised.

2.4.1 Parallel-sided channels

Consider first the flow in a channel with plane porous walls, $y = \pm h$, across which fluid is injected or extracted with constant uniform velocity V. This problem first attracted the attention of Berman (1953). With $\bar{x} = x/h, \quad \bar{y} = y/h$ the stream function is written as

$$\psi = h(U_0 - V\bar{x})f(\bar{y}), \tag{2.52}$$

where U_0 is an arbitrary velocity. With $\mu V/h$ as a scale for the pressure we have

$$\bar{p} = f' - \tfrac{1}{2}Rf^2 - \tfrac{1}{2}\lambda\bar{x}^2 + K\bar{x}, \tag{2.53}$$

where $R = Vh/\nu$ is the Reynolds number, positive for suction and negative for injection, λ is a parameter to be determined and $K = U_0\lambda/V$. We note that U_0 has no significant role to play, being simply the result of a constant pressure gradient along the channel. From (1.9), (2.52) and (2.53) we have, as the equation for f,

$$f''' + R(f'^2 - ff'') - \lambda = 0, \tag{2.54}$$

and if the flow is assumed to be symmetrical about the channel centre-line the boundary conditions are

$$f(0) = f''(0) = 0, \quad f(1) = 1, \quad f'(1) = 0.$$

That four conditions are required reflects the fact that λ is not determined *a priori*.

Berman (1953) constructed solutions of (2.54) for $|R| < 1$, whilst Yuan (1956) concentrated on the case $|R| \gg 1$. Terrill (1964, 1965) extended Berman's results, determined solutions numerically for $R = O(1)$, and on re-examining Yuan's results for $|R| \gg 1$ concluded they were valid only for $R < 0$, and then with anomalous behaviour at $\bar{y} = 0$ which he corrected, as well as considering the case $R \gg 1$. As $|R| \to \infty$ a boundary layer forms, either at $\bar{y} = 1$ with thickness $O(R^{-1})$ for $R > 0$, or at $\bar{y} = 0$ with thickness $O(|R|^{-1/2})$ for $R < 0$. Dual solutions were observed by Raithby (1971) for $R > 12$ but it was Robinson (1976) who uncovered the flow structure in greater detail. Thus, with $R > 0$ there are Type I solutions for which the velocity maximum is on the centre-line $\overline{y} = 0$, which are the solutions obtained by Terrill. For $R < 12.165$ these are the only solutions available, but when $R > 12.165$ two further solutions emerge. When $R > 13.119$ there are Type II solutions for which $f'(\overline{y}) > 0$, with the maximum velocity located between the centre-line and solid boundary. And finally there are Type III solutions. For $12.165 < R < 13.119$ there are dual solutions of this type, characterised by $f'(0) < 0$ and a region of reversed flow beyond which the velocity is positive. For $R > 13.119$ these Type III solutions continue to display the same characteristics. As $R \to \infty$, Type I and II solutions are virtually indistinguishable, differences being attributed to exponentially small terms. In figure 2.9 these features are illustrated.

Robinson confined his attention to values of $R > 0$, whereas Terrill's original (1964) investigation included values of $R < 0$, a continuation of the solutions now labelled Type I. It may be noted that Skalak and Wang (1978) have proved that there is at most one such solution whilst Shih (1987) has established the existence of a solution for all $R < 0$. In a penetrating analysis of the problem, Zaturska, Drazin and Banks (1988) have shown that when $R > 0$ bifurcations occur at $R = 6.001353$ on the Type I solution branch and at $R = 15.4146$ on the Type III solution branch. Each of these implies the existence of asymmetric solutions in the neighbourhood of each bifurcation point. Asymmetric solutions occur more naturally if the porous boundaries have different permeabilities, that is the suction and/or injection velocities are different at each boundary. In a series of papers Terrill and Shrestha (1965), and Shrestha and Terrill (1968), have considered such problems when there is suction or injection across each boundary, or suction at one and injection at the other, including the situation in which one wall is impermeable. Analyses for $R \ll 1$, $R \gg 1$ are complemented by numerical solutions.

Cox (1991a) continues the asymmetric solutions of the symmetric problem to derive new solutions of the asymmetric problem. In particular Cox (1991b) considers the situation in which one wall is impermeable. A generalisation of the problem originally posed by Berman, to three-dimensional flow, has been considered by Taylor, Banks, Zaturska and Drazin (1991).

In a problem, not unrelated to some of those considered in section 2.3, at least in the sense that stretching boundaries are involved, Brady and Acrivos (1981) consider the flow in a channel due to the stretching of its boundaries. Although this is an idealised problem the motivation was provided from the study of the flow within a long slender drop located in an extensional flow.

Specifically, Brady and Acrivos consider a channel $-h \leq y \leq h$ whose (impermeable) boundaries each move in their own plane with speed $U_0 x / h$.

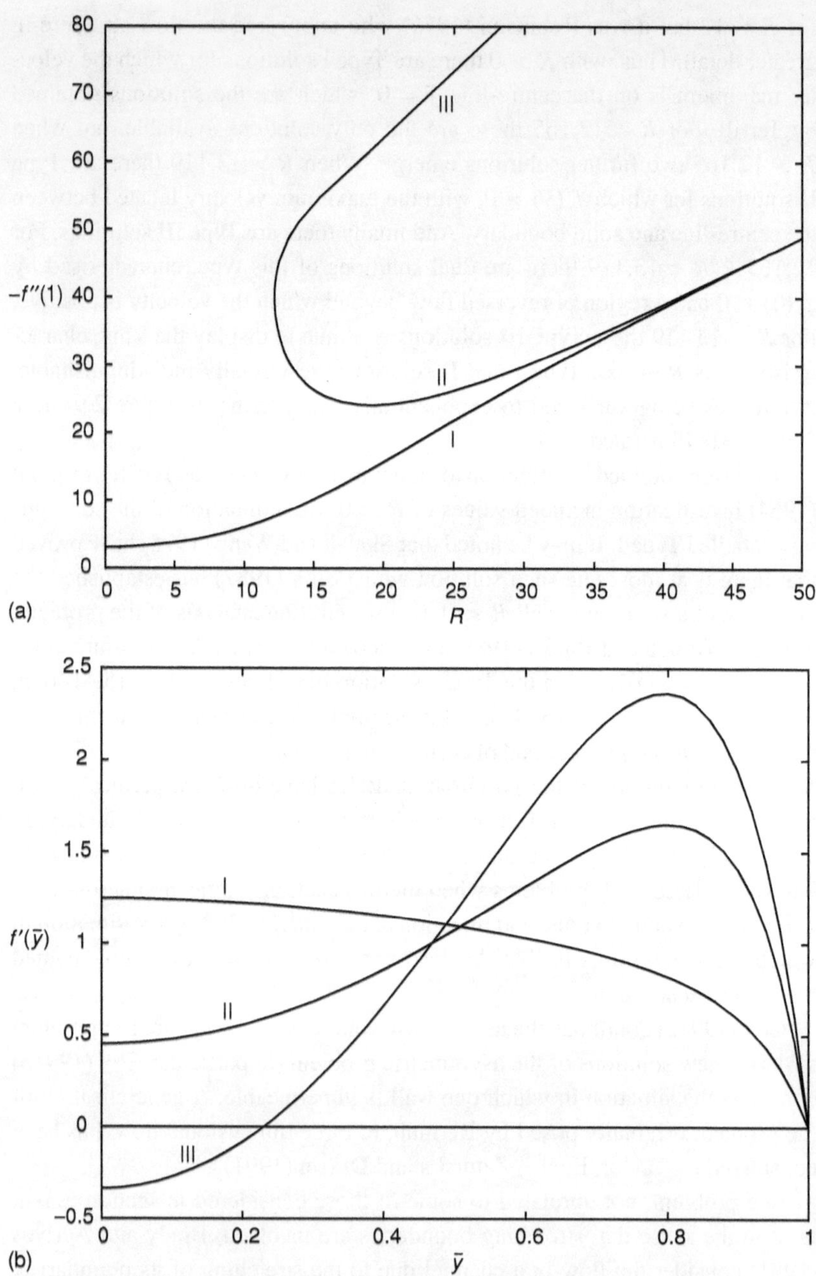

Figure 2.9 (a) Variation of $f''(1)$ with R for $R > 0$. (b) Velocity profiles for solutions of Type I with $R = 15.0141$, Type II with $R = 14.7151$, and Type III with $R = 12.5609$.

Since the flow is antisymmetric about $x = 0$, and symmetric about $y = 0$, only the flow in $x \geq 0$, $0 \leq y \leq h$ need be considered. Again we set $\bar{x} = x/h$, $\bar{y} = y/h$ and write

$$\psi = U_0 h \bar{x} f(\bar{y}), \tag{2.55}$$

which is essentially the Hiemenz transformation, and with $\mu U_0 / h$ as a scale for the pressure we have

$$\bar{p} = \tfrac{1}{2}\lambda \bar{x}^2 - \left(f' + \tfrac{1}{2}Rf^2 \right), \tag{2.56}$$

where $R = U_0 h / \nu$ is the Reynolds number and, again, λ is to be determined.

From (1.9), (2.55) and (2.56) we find, for $f(\bar{y})$

$$f''' + R(ff'' - f'^2) - \lambda = 0, \tag{2.57}$$

with boundary conditions

$$f(0) = f''(0) = 0, \quad f(1) = 0, \quad f'(1) = 1. \tag{2.58}$$

Numerical solutions of (2.57), subject to (2.58) have been obtained by Brady and Acrivos.

For $R \to 0$ we have $\lambda = 3$, $f = \bar{y}(\bar{y}^2 - 1)/2$; the flow is typical of what one might expect, namely with $f' > 0$ for $\bar{y}_0 < \bar{y} < 1$, where $\bar{y}_0 = 3^{-1/2}$ in this limit, with a reversed core flow for $0 < \bar{y} < \bar{y}_0$, and maximum flow reversal on the centre-line. This solution, designated Type I, varies continuously, with $\lambda \to -1$ as $R \to \infty$, displaying the same features throughout. For $R \gg 1$ the reversed core flow is essentially inviscid and uniform with velocity $O(U_0 R^{-1/2})$, flanked by boundary layers, with classical boundary-layer thickness $O(h R^{-1/2})$, at each boundary. The solution is unique for $R \leq R_c \approx 310$. Beyond $R = R_c$ there are three solutions corresponding to distinct values of λ. On an upper branch we have $\lambda \to -1$ as $R \to \infty$, and these so-called Type II solutions are almost indistinguishable from Type I. However as R decreases interesting new developments appear. The centre-line velocity at first increases, and then decreases until at $R = 337.4$ it vanishes. Meanwhile the maximum reversed core-flow velocity has established itself at $\bar{y} \approx 0.68$. Continuing along this branch, which has $\lambda \to -\infty$ as $R \to \infty$ with $|\lambda| = O(R)$ we have Type III solutions, with the positive centre-line velocity increasing, and the whole region of reversed flow now off the centre-line. Eventually, as $R \to \infty$, an inviscid core flanked by thin, classical boundary layers is again established. It may be noted that for $R > 100$ the shear stresses at $y = 1$ for the three solutions are indistinguishable numerically with $f''(1) = R^{1/2}$.

In this investigation Brady and Acrivos considered only the situation $R > 0$, corresponding to walls accelerating away from the origin. They advanced physical reasons for restricting their choice in this way. Subsequently Durlofsky and Brady (1984) found solutions of Type I for $R < 0$, for which case the walls accelerate towards the origin with outflow in the core. In a comprehensive study of flows of Type I, Watson, Banks, Zaturska and Drazin (1991) have discovered the existence of asymmetric solutions; these manifest themselves via pitchfork bifurcations at $R = 132.75849$ and $R = -17.30715$.

Cox (1991b) has extended the range of solutions to the case in which one wall accelerates whilst the other is at rest. While Watson *et al.* (1991) find additional solutions by combining porous and accelerating boundaries, of which the problem of Becker (1976) might be considered to be a special case.

2.4.2 Non-parallel-sided channels

One of the most celebrated exact solutions of the Navier–Stokes equations is associated with the flow between non-parallel plane walls of included angle 2α. The walls intersect at the origin where there is a source or sink leading to either a diverging or converging flow. The problem was first studied by Jeffery (1915) and Hamel (1916). Subsequently, special aspects of the problem have been addressed by Harrison (1919), von Kármán (1924), Tollmien (1931), Noether (1931) and Dean (1934). However, more comprehensive treatments have been given by Rosenhead (1940), Millsaps and Pohlhausen (1953), Berker (1963) and Whitham (1963). In what follows we adopt the approach of Whitham.

If Q is the volume flux per unit distance perpendicular to the plane of the flow, which may be either positive or negative, then, since it has dimensions L^2/T, $R = Q/\nu$ can be adopted as a Reynolds number. Further, since Q, ν are the only physical parameters of the problem then $r v_r/\nu$ and $r v_\theta/\nu$ must depend upon Q/ν and θ alone. With a similar argument applied to the pressure difference $p - p_0$ we have

$$v_r = \frac{\nu F(\theta)}{r}, \qquad v_\theta = \frac{\nu G(\theta)}{r}, \qquad \frac{p - p_0}{\rho} = \frac{\nu^2 P(\theta)}{r^2}. \qquad (2.59)$$

From the continuity equation (1.22) we have $G(\theta) = \text{constant} \equiv 0$ since $v_\theta = 0$ at the boundaries $\theta = \pm\alpha$. Substitution of (2.59) into (1.19) and (1.20) then yields $P' = 2F'$ so that $P = 2F + C_1$, and $P = -(F^2 + F'')/2$. It follows that F satisfies

$$F'' + F^2 + 4F + 2C_1 = 0,$$

from which, multiplying by F' and integrating, we have

$$\tfrac{1}{2}F'^2 + \tfrac{1}{3}F^3 + 2C_1 F = C_2,$$

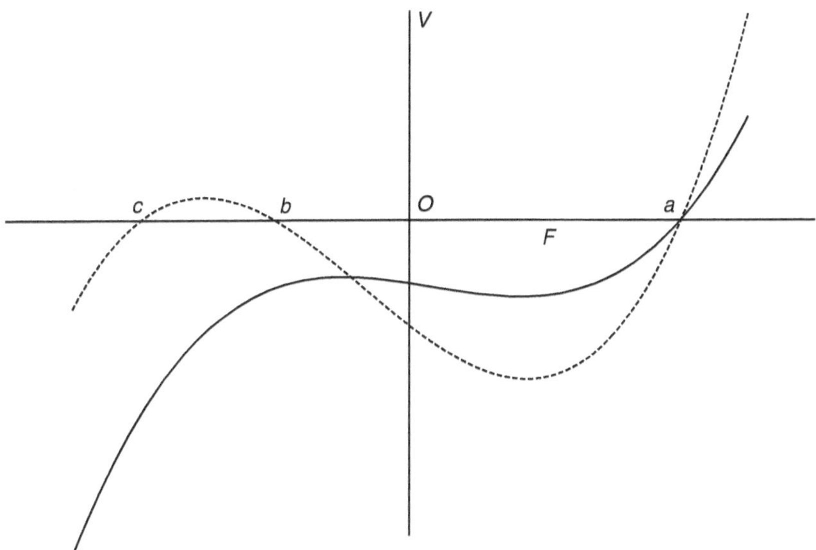

Figure 2.10 A sketch of the two possible forms for $V(F)$.

which in turn may be written as

$$\tfrac{1}{2}F'^2 - \tfrac{1}{3}(a - F)(F - b)(F - c) = 0, \tag{2.60}$$

where the constants a, b, c are related to C_1, C_2, and $a + b + c = -6$. The range and type of solutions available, for which the no-slip condition requires $F(\pm\alpha) = 0$, may be illustrated by a dynamical interpretation of equation (2.60). Thus, with θ as 'time' and F as 'displacement' we may interpret (2.60) as the energy equation for a particle of unit mass, moving on a straight line, whose potential energy is

$$V(F) = -\tfrac{1}{3}(a - F)(F - b)(F - c). \tag{2.61}$$

The total energy is zero and $V \leq 0$ in the motion. Since the particle starts at $F = 0$ for $\theta = -\alpha$, and returns to $F = 0$ for $\theta = \alpha$, there are essentially two situations to consider. The first has b, c complex conjugates with $a > 0$, whilst for the second all are real with $c < b < 0 < a$. These are illustrated in figure 2.10. In the first case the particle starts at $F = 0$ with finite positive velocity, reaches $F = a$ where its velocity is zero but the acceleration $F'' = -\mathrm{d}V/\mathrm{d}F$ is negative, and returns to 0. The motion beyond 0 is of no concern and the particle moves off to infinity. The particle motion in $0 < F < a$ describes a fluid motion that is a pure outflow, since $F > 0$, and symmetric about $\theta = 0$. For the second case the situation is more complex. A motion of the particle from $F = 0$ to $F = a$ returning to $F = 0$ again interprets as symmetric pure outflow. Correspondingly a motion from $F = 0$ to $F = b$ returning to $F = 0$, with $F < 0$ throughout, interprets as symmetric pure inflow. But now, of course, the particle may oscillate in $b \leq F \leq a$ corresponding to a fluid flow which has both regions of outflow and regions of inflow, with the flow not necessarily symmetric about $\theta = 0$. For each outflow or inflow region the extrema of F are a or b respectively. This rich structure of solutions is summarised by Rosenhead (1940) from whom we quote:

> For every pair of values of α and R the number of mathematically possible velocity profiles of radial motion is infinite.... The profiles may or may not be symmetrical with respect to the central line of the channel. If $\pi > \alpha > \pi/2$ *pure* outflow is impossible, and there is a range of values of small Reynolds number in which *pure* inflow is impossible. The effect of increasing R in outflow is to exclude, progressively, more and more of the simpler types of flow. No such exclusion is introduced when $|R|$ is increased in inflow. With increasing Reynolds number in pure inflow, and with small values of α, the velocity profile exhibits all the well-known characteristics of boundary layers near the walls, and an approximately constant velocity across the rest of the channel. In pure outflow the flow becomes more and more concentrated in the centre of the channel as R is increased, until finally regions of inflow occur near the wall.

In the above context the 'simplest' type of flow is pure outflow or inflow. For pure outflow we have, since $F = a$ when $\theta = 0$, from equation (2.60)

$$\theta = \pm\sqrt{\frac{3}{2}} \int_F^a \frac{\mathrm{d}t}{\sqrt{(a-t)(t-b)(t-c)}}, \tag{2.62}$$

from which we infer that

$$\alpha = \sqrt{\frac{3}{2}} \int_0^a \frac{\mathrm{d}F}{\sqrt{(a-F)(F-b)(F-c)}}, \tag{2.63}$$

and we have

$$\frac{1}{2}R = \sqrt{\frac{3}{2}} \int_0^a \frac{F\,\mathrm{d}F}{\sqrt{(a-F)(F-b)(F-c)}}. \tag{2.64}$$

For pure inflow the limits of integration are from b to F, or 0, in the above. As with the heuristic discussion associated with the dynamics of a single particle, more complicated flows may be deduced from appropriate combinations of these. For given values of α and R the constants a, b, c may, in principle, be determined from (2.63), (2.64) and the constraint $a + b + c = -6$, and the solution completed from (2.62).

Whilst computational methods can complete the solution it is possible to express it in terms of tabulated functions by a suitable transformation. Thus in the first case with a real and b, c complex conjugates, only pure outflow is possible with

$$F(\theta) = a - \frac{3M^2}{2} \frac{1 - \mathrm{cn}(M\theta, \kappa)}{1 + \mathrm{cn}(M\theta, \kappa)}, \tag{2.65}$$

where $M^2 = 2\{(a-b)(a-c)\}^{1/2}/3$, $2\kappa^2 = 1 + (a+2)/M^2$. For the second case with a, b, c real we have, for outflow,

$$F(\theta) = a - 6k^2 m^2 \, \mathrm{sn}^2(m\theta, k), \tag{2.66}$$

and for inflow

$$F(\theta) = a - 6k^2 m^2 \, \mathrm{sn}^2\{K(k^2) - m\theta, k\}, \tag{2.67}$$

where $m^2 = (a-c)/6, k^2 = (a-b)/(a-c)$ and K is the first complete elliptic integral.

The limiting form for pure outflow will have $F'(\pm\alpha) = 0$ which implies, from (2.60), that $b = 0$. In that case we see from (2.63) that the critical value of α, say α_c, is

$$\begin{aligned}
\alpha_c &= \sqrt{\frac{3}{2}} \int_0^a \frac{\mathrm{d}F}{\sqrt{F(a-F)(F+a+6)}} \\
&= \sqrt{\frac{3}{2a}} \int_0^1 \frac{\mathrm{d}t}{\sqrt{t(1-t)\{1+(1+6/a)t\}}} = \sqrt{\frac{6}{a}} k K(k^2),
\end{aligned} \tag{2.68}$$

with $k^2 = a/\{2(3+a)\}$ in this limiting case, which determines $\alpha_c = \alpha_c(a)$. One may note that $a = rv_r|_{\max}/\nu$ may be interpreted as a Reynolds number of the flow, and is used as such by Millsaps and Pohlhausen (1953). The corresponding critical flux

$$R_c = \frac{Q_c}{\nu} = 2 \int_0^{\alpha_c} \{a - 6k^2 m^2 \, \mathrm{sn}^2(m\theta, k)\} \, \mathrm{d}\theta = 12 \frac{\{E(k^2) - (1 - k^2)K(k^2)\}}{\sqrt{(1 - 2k^2)}}, \tag{2.69}$$

where E is the second complete elliptic integral. Since the simplest type of flow is pure inflow or outflow, the critical curve $\alpha_c = \alpha_c(R)$ which places a limit on such flows is of particular importance and is shown in figure 2.11. For $a \gg 1$, $k^2 \sim (1 - 3/a)/2$ and from (2.68), (2.69) we have $\alpha_c \sim 9.42/R$.

For pure inflow, given by (2.67), as $-b$ increases the flow tends to become uniform, except for boundary layers close to the walls. To see this note that m^2 is large and so for given α, bounding θ, K must be large and so $k \approx 1$, $\mathrm{sn}\,t \approx \tanh t, c \approx b, a \approx -2b$ and we have

$$F(\theta) = b[3 \tanh^2\{(-b/2)^{1/2}(\alpha - \theta) + \beta\} - 2],$$

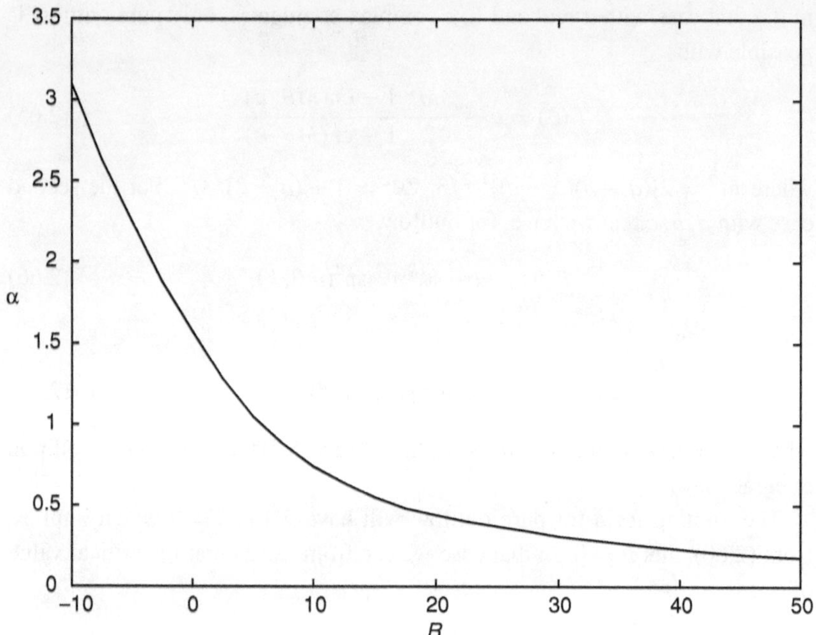

Figure 2.11 The critical curve $\alpha_c = \alpha_c(R)$. Below this pure inflow or outflow is possible, above it regions of reversed flow must be present.

where $\beta = \tanh^{-1}\sqrt{2/3}$, from which we see that $F \approx b$ except in boundary layers of thickness $O\{(-b)^{-1/2}\}$. The volume flux is given by $Q/\nu \approx 2\alpha b$ so that $|R| = O(|b|)$, and the boundary layers have classical thickness $O(|R|^{-1/2})$. In figure 2.12 we show velocity profiles for both diverging and converging flows. These have been obtained by direct numerical integration of the differential equation for F, utilising data set out in Millsaps and Pohlhausen (1953).

The formal requirement that there is a line source, or sink, at the intersection of the plane boundaries appears to limit the usefulness of the Jeffery–Hamel flows. However, Fraenkel (1962, 1963) has demonstrated their applicability to flows bounded by slightly curved walls, originating from a channel of non-vanishing width. The walls are required to turn smoothly, and the 'slightly curved' condition demands that

$$\kappa = \text{local wall curvature} \times \text{local channel width} \ll 1.$$

There is no restriction on the local angle α the wall makes with the channel centre -line. With these requirements satisfied, Fraenkel (1962) shows that a Jeffery–Hamel solution based on the local divergence angle, α, and volume flow through the channel, Q, provides a first approximation to the exact solution at every

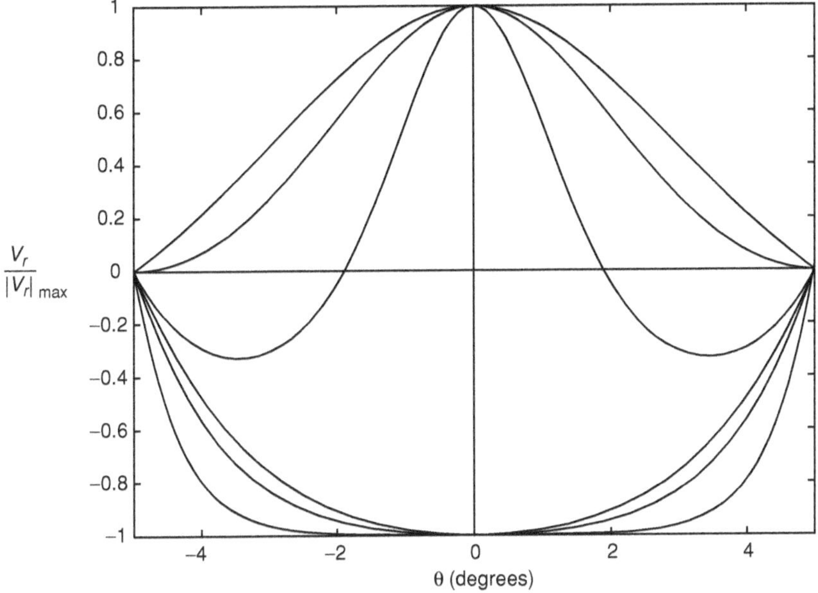

Figure 2.12 Velocity profiles for inflow and outflow in a channel of included angle $10°$. If a Reynolds number $R_m = r v_r|_{\max}/\nu$ is defined then, from the bottom, the profiles correspond to $R_m = -5000, -1342, -684, 5000, 1342, 684$, respectively.

station $r =$ constant, With this as the leading term Fraenkel (1963) develops the solution in series form. One aim of these investigations is the construction of solutions of the Navier–Stokes equations, in channels of varying width, which involve flow separation.

Generalisations of the classical case have been considered when a normal velocity proportional to distance from the vertex is imposed on both boundaries (Moffatt and Duffy (1980)), and to the case in which the viscosity and density are arbitrary functions of θ (Hooper, Duffy and Moffatt (1982)).

In Hamel's (1916) investigation the flow between non-parallel plane walls emerged as a special case of spiral flow between solid walls, where the results are similar to those obtained above. In his study of spiral flows he showed, in particular, that streamlines in the form of equiangular spirals are the only streamlines which can coincide with the streamlines of a potential flow without the motion itself being irrotational. An extensive account of the work by Hamel may be found in Berker (1963).

In a related problem Berker (1963), and most recently Putkaradze and Dimon (2000), consider flows of the form (2.59), (2.60) in the absence of any boundaries. Such a flow corresponds to a non-uniform two-dimensional source. In particular Putkaradze and Dimon, seeking solutions with k-fold symmetry with respect to θ, show that solutions only exist for $R \leq \pi(k^2 - 4)$, a result also given by Goldshtik and Shtern (1989). Putkaradze (2003) also considers the radial flow, from a two-dimensional source, of two immiscible fluids separated by radial lines. In that case the boundary conditions of the classical flow are replaced by conditions of continuity of velocity and stress at the fluid–fluid interfaces.

2.5 Three-dimensional flows

In this section we consider flows that take place not solely in the (x, y)-plane, but also have a component of velocity in the z-direction.

2.5.1 A corner flow

Stuart (1966b) considers flows for which the velocity components $u = \phi_x$, $v = \phi_y$ where $\phi = \phi(x, y)$ is a velocity potential with $\nabla^2 \phi = 0$, and $w = w(x, y)$. With the pressure p given by $p/\rho = -(\phi_x^2 + \phi_y^2)/2$, equations (1.9) (1.10) are, in the absence of any body force, satisfied identically and, from (1.11), w satisfies

$$\frac{\partial \phi}{\partial x} \frac{\partial w}{\partial x} + \frac{\partial \phi}{\partial y} \frac{\partial w}{\partial y} = \nu \left(\frac{\partial^2 w}{\partial x^2} + \frac{\partial^2 w}{\partial y^2} \right),$$

with solution

$$w = A + B \, e^{\phi/\nu}. \tag{2.70}$$

The solution is completed when ϕ is determined. However, note that if $w = 0$ at some solid surface, then at that surface $\phi = \nu \ln(-A/B)$, but if $\phi = $ constant at the surface then the normal derivative $\partial \phi / \partial n \neq 0$ there, in general. This implies that the solution involves a porous surface at which suction is applied. If $\phi \to -\infty$ away from any solid boundary then $w \to A = W_0$, say. A particularly simple example has $\phi = -Vy$ and with $w = 0$ at $y = 0$ the asymptotic suction profile $w = W_0(1 - e^{-Vy/\nu})$ is recovered. As an extension of this Stuart takes $\phi = -kxy$ so that with

$$u = -ky, \quad v = -kx, \quad w = W_0 \left(1 - e^{-kxy/\nu}\right),$$

which corresponds to uniform flow along the corner $x = 0, y \geq 0; y = 0, x \geq 0$. To sustain this flow the suction must increase linearly with distance from the corner.

Stuart also shows how this solution may be extended to the flow in a corner of any included angle less than 2π.

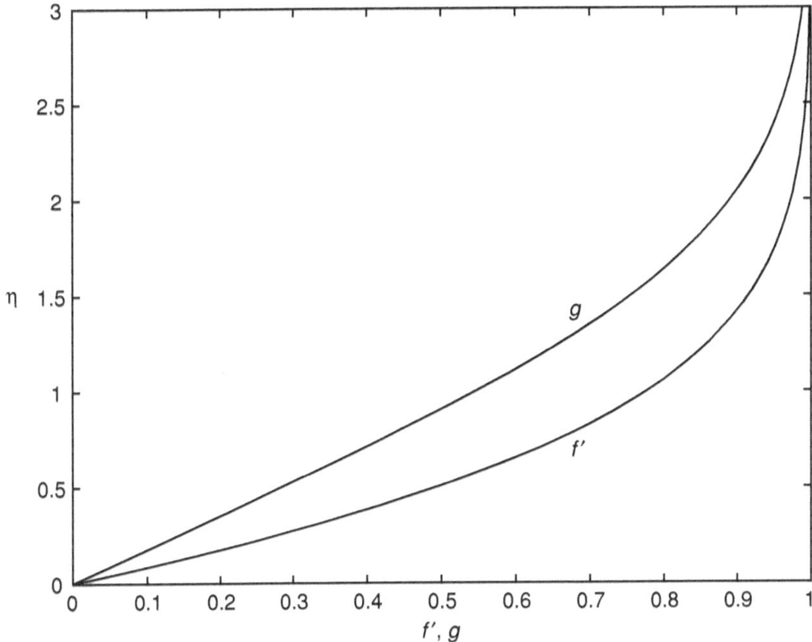

Figure 2.13 The Hiemenz velocity profile $f'(\eta)$ and the cross-flow velocity profile $g(\eta)$ at a swept stagnation line.

2.5.2 A swept stagnation flow

In section 2.3 we considered the classical stagnation-point flow appropriate, for example, to the flow at the stagnation line of a cylinder placed perpendicular to an oncoming stream. If the cylinder is 'swept', or placed at an angle to the stream, then (2.32) is supplemented by taking the third velocity component as $w = w_0 g(\eta)$ and from (1.11) and (2.32) we find

$$g'' + fg' = 0, \quad \text{with} \quad g(0) = 0, \quad g(\infty) = 1,$$

so that

$$g(\eta) = \frac{\int_0^\eta \exp\left(-\int_0^t f \, ds\right) dt}{\int_0^\infty \exp\left(-\int_0^t f \, ds\right) dt}. \tag{2.71}$$

The velocity profile $w/w_0 = g(\eta)$ is shown in figure 2.13.

Coward and Hall (1996) extend this solution to include a thin liquid film at the surface of the cylinder.

2.5.3 Vortices in a stagnation flow

Kerr and Dold (1994) have considered a flow in which, superposed upon a two-dimensional stagnation flow in the absence of any solid boundaries, is an array of counter-rotating vortices whose axes are parallel to the direction of the diverging flow. The velocity field may be written as

$$\left(\frac{A}{k}x, \quad -\frac{A}{k}y + \nu k \frac{\partial \psi}{\partial z}, \quad -\nu k \frac{\partial \psi}{\partial y} \right), \tag{2.72}$$

where A is a measure of the strength of the basic stagnation flow, $\psi = \psi(y, z)$ is a stream function made dimensionless with scale ν, and k^{-1} is a length chosen such that $2\pi/k$ is the periodic spacing in the z-direction of the vortices. The vorticity

$$\boldsymbol{\omega} = \xi \mathbf{i} = -\left(\frac{\partial^2 \psi}{\partial y^2} + \frac{\partial^2 \psi}{\partial z^2} \right) \mathbf{i}, \tag{2.73}$$

and the equation satisfied by the vorticity is

$$-\frac{\partial \psi}{\partial y}\frac{\partial \xi}{\partial z} + \frac{\partial \psi}{\partial z}\frac{\partial \xi}{\partial y} - \lambda y \frac{\partial \xi}{\partial y} - \lambda \xi = \frac{\partial^2 \xi}{\partial y^2} + \frac{\partial^2 \xi}{\partial z^2}, \tag{2.74}$$

where $\lambda = A/\nu k^2$, which may be interpreted as a measure of the strength of the converging flow to the rate of viscous diffusion.

A simple solution of the above is, for $\lambda = 1$, $\psi = \xi = \cos z$, obtained by Craik and Criminale (1986) as a special case in their examination of the stability of disturbances that consist of single Fourier modes in unbounded shear flows. For the vortex flows Kerr and Dold develop solutions of (2.73), (2.74) as

$$\xi = \sum_{n=1}^{\infty} a_n(y) \cos(2n-1)z + b_n(y) \sin 2nz,$$

$$\psi(y, z) = \sum_{n=1}^{\infty} c_n(y) \cos(2n-1)z + d_n(y) \sin 2nz.$$

Substitution into the equations for ψ, ξ yields an infinite system of ordinary differential equations, namely

$$\begin{aligned}
a_n'' + \lambda y a_n' + \{\lambda - (2n-1)^2\}a_n &= F_{2n-1}(\psi, \xi), \\
b_n'' + \lambda y b_n' + (\lambda - 4n^2)b_n &= F_{2n}(\psi, \xi), \\
c_n'' + (2n-1)^2 c_n &= -a_n, \\
d_n'' - 4n^2 d_n &= -b_n,
\end{aligned} \tag{2.75}$$

where F_{2n-1}, F_{2n} are the appropriate components of the non-linear terms. By symmetry we anticipate $\psi(y, z) = \psi(-y, -z) = -\psi(y, \pi - z)$ so that $a_n(y)$,

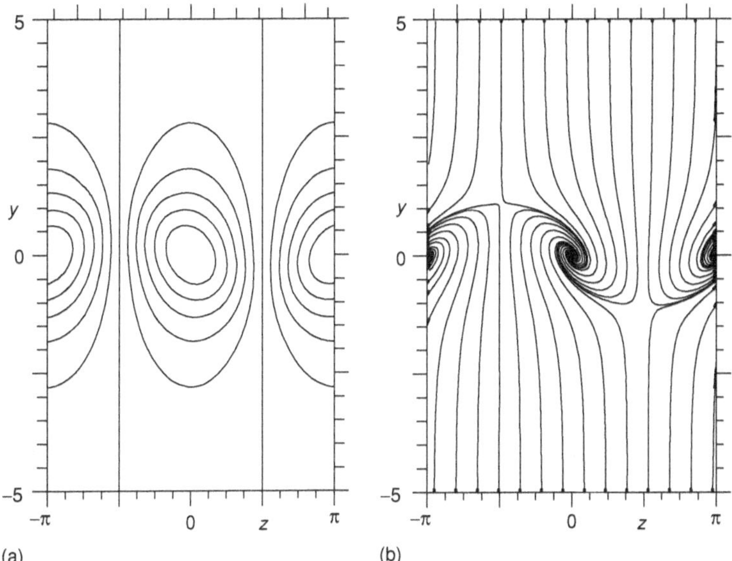

Figure 2.14 (a) Streamlines of perturbation vortices of amplitude 10 for $\lambda = 6$. (b) The corresponding streamlines for the whole flow projected onto the (y, z)-plane. (From Kerr and Dold (1994).)

$b_n(y)$ are even and odd functions of y respectively. Thus

$$a'_n(0) = b_n(0) = c'_n(0) = d_n(0) = 0,$$

and in addition we require $a_n, b_n, c_n, d_n \to 0$ as $y \to \infty$. Decay of the solutions in the far field is ensured if $\lambda > 1$; accordingly there are essentially two parameters that characterise these flows, namely λ and a second parameter that Kerr and Dold take as the amplitude of the vortices, defined as the value of the stream function at the origin. In carrying out numerical solutions of (2.75) fewer than ten terms of the series are required for a wide range of values of λ and the amplitude. An example is shown in figure 2.14.

Kerr and Dold note that the vortex structures vary only slowly as the amplitude increases from infinitesimal to large values, and that these are mirrored, to some extent, when the non-linear terms in (2.75) are ignored. That this is so may be expected, since the essential balance between diffusion of vorticity and intensification due to stretching of vortex lines is associated with the linear terms in (2.74).

Kerr and Dold note that an array of counter-rotating vortices are observed in the 'four roll mill' experiments of Lagnado and Leal (1990). The vortices appear above a critical strain rate with axes aligned with the straining flow.

Burgers (1948) and Robinson and Saffman (1984) have considered a simpler configuration, namely a shear flow in the z-direction, that nevertheless retains the essential balance between diffusion of vorticity and intensification due to stretching. Thus, with $\mathbf{v} = \{\alpha x, -\alpha y, W(y)\}$, we have $\boldsymbol{\omega} = \{\xi(y), 0, 0\}$, where $\xi = W'$ satisfies $\nu\xi'' + \alpha(y\xi)' = 0$ with solution

$$\xi = \{W(\infty) - W(-\infty)\} \left(\frac{\alpha}{2\pi\nu}\right)^{1/2} \exp(-\alpha y^2/2\nu), \tag{2.76}$$

and $W(\infty) - W(-\infty)$ is a free parameter.

2.5.4 Three-dimensional stagnation-point flow

In section 2.3 above we have considered the flow at a two-dimensional stagnation point on a plane boundary, and in the next chapter we shall discuss its axisymmetric analogue. Both of these are special cases of a three-dimensional stagnation-point flow on a plane boundary, first considered by Howarth (1951). For consistency with the established literature, and with chapter 3, we adopt a slight change of notation so that x, y are co-ordinates in the plane, with z perpendicular to it. With $u = kx$, $v = ly$ at large distances from the boundary it proves convenient to write

$$u = kxf'(\eta), \quad v = kyg'(\eta), \quad w = -(\nu k)^{1/2}\{f(\eta) + g(\eta)\}, \quad \eta = \left(\frac{k}{\nu}\right)^{1/2} z, \tag{2.77}$$

so that, with $r = l/k$, substitution into equations (1.9) to (1.11) gives, as equations for f and g,

$$f''' + (f + g)f'' + 1 - f'^2 = 0, \quad g''' + (f + g)g'' + r^2 - g'^2 = 0,$$

with the pressure determined as

$$\frac{p_0 - p}{\rho} = \frac{1}{2}(k^2 x^2 + l^2 y^2) + \nu k(f' + g') + \frac{1}{2}\nu k(f + g)^2.$$

The boundary conditions require

$$f(0) = g(0) = f'(0) = g'(0) = 0, \quad f'(\infty) = 1, \quad g'(\infty) = r.$$

Howarth obtained solutions for f and g for a range of values of r with $0 \le r \le 1$. These solutions correspond to the flow in the immediate neighbourhood of a nodal point of attachment as might be found, for example, at an asymmetric protuberance on a body placed in a uniform stream parallel to the z-direction. Subsequently Davey (1961) obtained solutions for a range of negative values of r with $-1 \le r < 0$. Such solutions correspond to a saddle point of

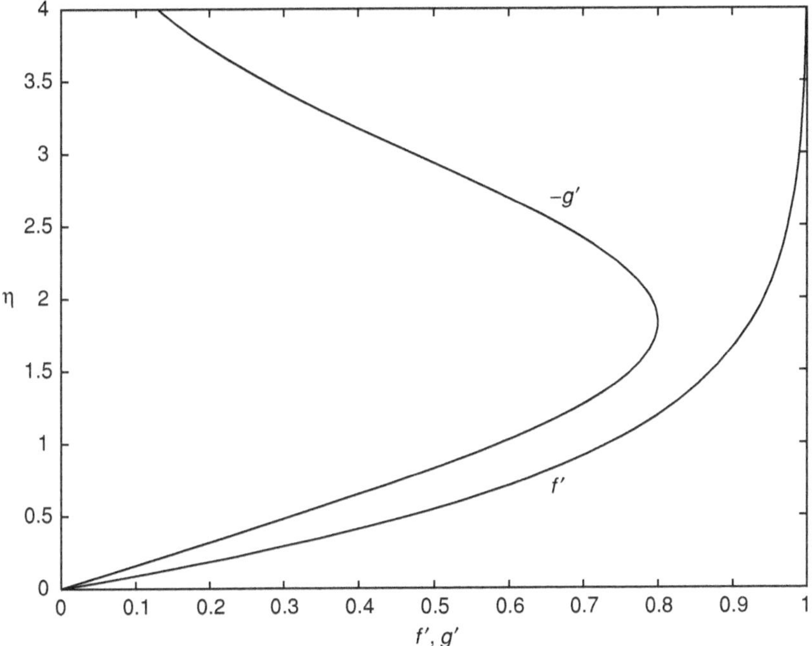

Figure 2.15 Velocity profiles for the dual of the two-dimensional stagnation-point flow (i.e. $r = 0$) obtained by Davey and Schofield (1967).

attachment which may be found, for example, in the immediate neighbourhood of a geometrical saddle point on the surface of a body located between two protuberances. From a topological viewpoint Davey establishes that on a finite body the number of nodal points must exceed the number of saddle points by two. He also proves that for $r < -1$ which, as we see from (2.77), corresponds to a saddle point of separation, the equations have no solution.

Subsequently Libby (1967), Davey and Schofield (1967) and Schofield and Davey (1967) revisited the problem and demonstrated the existence of dual solutions of (2.77). Libby suggested non-uniqueness for $r < 0$ and gave some examples. Schofield and Davey, however, demonstrated the existence of a dual for each of the solutions obtained by Howarth (1951) and Davey (1961) in the range $-1 \le r \le 1$. The particular case $r = 0$ was discussed in detail by Davey and Schofield. This corresponds to a three-dimensional solution of the Navier–Stokes equations embedded in what is essentially a two-dimensional stagnation point of attachment. Solutions were obtained numerically by Davey and Schofield and we show the velocity profiles $f'(\eta)$, $g'(\eta)$ in figure 2.15. Similarly, when $r = 1$, the dual of the axisymmetric problem is asymmetric.

Some years later Hewitt, Duck and Stow (2002) showed that these two solution branches, for which $1 - f', r - g'$ are exponentially small as $\eta \to \infty$, are just the first two of a countably infinite sequence of such states.

Libby (1974) has extended the analysis of Rott (1956) for a two-dimensional stagnation point at a sliding boundary to the three-dimensional case. Thus, the velocity components u, v in equation (2.77) are replaced by

$$u = u_w F(\eta) + kx f'(\eta), \quad v = v_w G(\eta) + ly g'(\eta),$$

and the ordinary differential equations satisfied by F and G integrated numerically. The constants u_w, v_w represent the velocity of the boundary. This solution finds application, for example, to the flow at a stagnation point on an ellipsoid rotating about one of its axes.

3

Steady axisymmetric and related flows

3.1 Circular pipe flow

We begin this chapter with a discussion of the flow within a circular pipe, including the flow in the annulus between concentric circular pipes. With the flow driven along the pipe by a uniform pressure gradient we have the analogue of plane Poiseuille flow. In the absence of any pressure gradient a flow may be induced by the motion of one cylinder relative to another, either by a sliding motion parallel to the cylinder generators or by relative rotation about their common axis, which may be considered the analogue of plane Couette flow.

If the velocity components are assumed to be independent of θ and z, that is $v_r = v_r(r)$, $v_\theta = v_\theta(r)$ and $v_z = v_z(r)$, the continuity equation (1.22) shows that

$$v_r = \frac{m}{r}. \tag{3.1}$$

The momentum equations (1.19) to (1.21) then simplify to

$$v_r \frac{\partial v_r}{\partial r} - \frac{v_\theta^2}{r} = -\frac{1}{\rho} \frac{\partial p}{\partial r}, \tag{3.2}$$

$$v_r \frac{\partial v_\theta}{\partial r} + \frac{v_r v_\theta}{r} = \nu \left(\frac{\partial^2 v_\theta}{\partial r^2} + \frac{1}{r} \frac{\partial v_\theta}{\partial r} - \frac{v_\theta}{r^2} \right), \tag{3.3}$$

$$v_r \frac{\partial v_z}{\partial r} = -\frac{1}{\rho} \frac{\partial p}{\partial z} + \nu \left(\frac{\partial^2 v_z}{\partial r^2} + \frac{1}{r} \frac{\partial v_z}{\partial r} \right), \tag{3.4}$$

where v_r is given by (3.1). Under the assumptions made, the case $m \neq 0$ is only possible if the cylinder boundaries, at $r = a, b$ ($b < a$) are porous and the normal velocities there prescribed as m/a, m/b.

For the flows anticipated above, other boundary conditions to be applied are

$$v_\theta = a\Omega_a \quad \text{at} \quad r = a, \quad v_\theta = b\Omega_b \quad \text{at} \quad r = b, \tag{3.5}$$

45

and

$$v_z = 0 \quad \text{at} \quad r = a, \quad v_z = W \quad \text{at} \quad r = b. \tag{3.6}$$

With the pressure field given as $p = -Pz + \tilde{P}(r)$, where P is a constant, equations (3.2) to (3.4), subject to (3.5), (3.6) may readily be solved for v_θ, v_z and $\tilde{P}$. However for convenience, and historical reasons, the cases of axial motion and circular motion are treated separately.

(i) Case $m = \Omega_a = \Omega_b = 0$ For this case both v_r, v_θ vanish, and the solution of equation (3.4), subject to (3.6), is

$$v_z = \frac{P}{4\mu}(a^2 - r^2) + \left\{ W + \frac{P}{4\mu}(b^2 - a^2) \right\} \frac{\ln(r/a)}{\ln(b/a)} \tag{3.7}$$

with the corresponding volume flux along the annulus given by

$$Q = \frac{\pi P}{8\mu} \left\{ a^4 - b^4 + \frac{(a^2 - b^2)^2}{\ln(b/a)} - \frac{8\mu W b^2}{P} - \frac{4\mu W}{P\ln(b/a)}(a^2 - b^2) \right\}. \tag{3.8}$$

The particular case with $W = 0$, when the flow is driven along the pipe by the pressure gradient alone, was first considered by Boussinesq (1868). The special case of (3.7), (3.8)

$$W = \frac{Pb^2 \ln(b/a)}{2\mu} + \frac{P}{4\mu}(a^2 - b^2),$$

with

$$v_z = \frac{P}{4\mu}(a^2 - r^2) + \frac{Pb^2}{2\mu}\ln(r/a) \tag{3.9}$$

and

$$Q = \frac{\pi P}{8\mu} \left\{ a^4 - b^4 + 4b^4 \ln(a/b) - 4b^2(a^2 - b^2) \right\} \tag{3.10}$$

has $\partial v_z/\partial r = 0$ at $r = b$. As a consequence it is possible, in equations (3.9) and (3.10), to allow $b \to 0$ and recover the classical solution for flow down a pipe of radius a under the action of a constant pressure gradient,

$$v_z = \frac{P}{4\mu}(a^2 - r^2), \quad \text{with} \quad Q = \frac{\pi P a^4}{8\mu}.$$

This famous result for the flux was first found empirically by Hagen (1839) and Poiseuille (1840), who independently measured the flow of water along capillary tubes. Stokes (1845) first solved the equations of motion to find (essentially) these results for pipe flow, in agreement with the measurements of Hagen and Poiseuille.

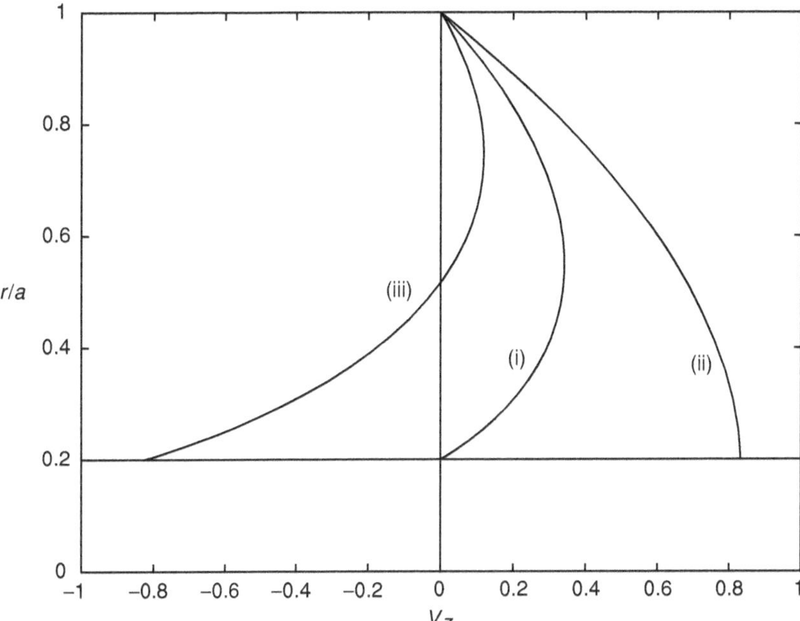

Figure 3.1 Annular pipe flow with non-zero axial pressure gradient P and $b/a = 0.2$. (i) The case $W = 0$, (ii) $W > 0$ with v_z as in equation (3.9), (iii) $W < 0$ such that $Q \equiv 0$.

The special case

$$\frac{P}{8\mu} = W\left[\frac{2b^2\ln(b/a) + a^2 - b^2}{2(a^2 - b^2)\{(a^2 + b^2)\ln(b/a) + a^2 - b^2\}}\right]$$

ensures $Q \equiv 0$. This finds application when the outer cylinder is of finite length L, and sealed at each end with the inner cylinder drawn through the seals, first discussed by Bouasse (see Berker (1963)). If $L \gg a$ the fully developed flow, (3.7), may be expected to be appropriate away from the ends.

Representative velocity profiles associated with these pipe flows are shown in figure 3.1.

(ii) Case $P = 0$ For this case, provided also $W = 0$, the flow may be driven by the rotation of the cylindrical boundaries which, for $m = 0$, may be expected to result in concentric circular streamlines or, if $m \neq 0$ corresponding to a source-like flow within the porous inner cylinder, spiral streamlines. The solution of equation (3.3) with conditions (3.5) is, with $v_r = m/r$,

$$v_\theta = \frac{a^2 b^2 (a^R \Omega_b - b^R \Omega_a)}{a^{R+2} - b^{R+2}}\frac{1}{r} + \frac{a^2 \Omega_a - b^2 \Omega_b}{a^{R+2} - b^{R+2}} r^{R+1} \qquad (3.11)$$

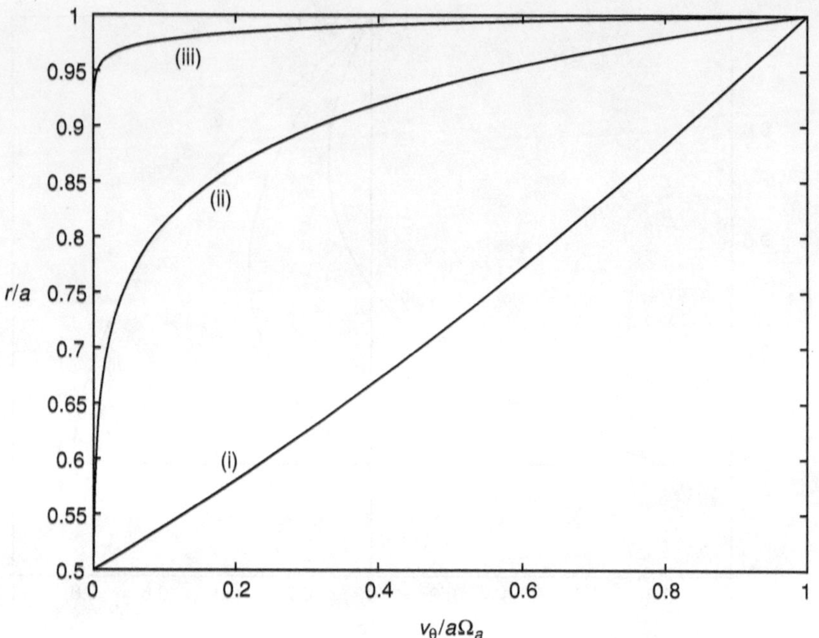

Figure 3.2 Annular pipe flow with $P = W \equiv 0$ and $b/a = 0.5$. The case $\Omega_b = 0$ is illustrated with (i) $R = 0$, (ii) $R = 10$, (iii) $R = 100$. As R increases the boundary-layer nature of the flow is apparent.

where $R = m/\nu$ is a Reynolds number or, in the exceptional case $R = -2$

$$v_\theta = \frac{b^2\Omega_b \ln a - a^2\Omega_a \ln b}{\ln(a/b)}\frac{1}{r} + \frac{a^2\Omega_a - b^2\Omega_b}{\ln(a/b)}\frac{\ln r}{r}.$$

The pressure function $\tilde{P}(r)$ follows from equation (3.2). These solutions were obtained by Hocking (1963), who commented on the case of high Reynolds number, $R \gg 1$, for which, from (3.11)

$$v_\theta \simeq \frac{b^2\Omega_b}{r} + \{a\Omega_a - (b^2/a)\Omega_b\}\left(\frac{r}{a}\right)^{R+1}.$$

It is clear that in this case, with the second term negligible unless $r \approx a$, the azimuthal velocity behaves as a potential vortex. In other words fluid particles issuing from the inner cylinder retain their initial angular momentum until the viscous torque close to $r = a$ changes it. As a consequence the flow divides into an essentially inviscid potential core with a thin boundary layer close to $r = a$, whose thickness may be estimated as $O(aR^{-1})$. Profiles of the velocity component v_θ for various values of R in the special case $\Omega_b = 0$ are shown in figure 3.2.

For $R = 0$, corresponding to impermeable boundaries, the only non-zero component of velocity is v_θ given, from equation (3.11), as

$$v_\theta = \frac{a^2 b^2 (\Omega_b - \Omega_a)}{a^2 - b^2} \frac{1}{r} + \frac{a^2 \Omega_a - b^2 \Omega_b}{a^2 - b^2} r,$$

a solution whose form was known to Stokes (1845). For this case the moment per unit length of each fluid cylinder of radius r is

$$M = -4\pi \mu a^2 b^2 \left(\frac{\Omega_a - \Omega_b}{b^2 - a^2} \right).$$

Viscometers based upon Poiseuille flow or concentric rotating cylinders have been used for determining the viscosity of fluids, Mallock (1896).

A steady flow external to the cylinders can, in general, only be sustained if suction is applied at the cylinder. Consider a uniform axial flow $(0, 0, W)$ external to the porous cylinder $r = a$. Stuart (1966b) has shown that the appropriate solution is

$$v_r = -\frac{m}{r}, \quad v_z = W \left\{ 1 - \left(\frac{a}{r} \right)^R \right\}, \quad R = \frac{m}{\nu},$$

noting that for $R \gg 1$ a boundary layer forms at $r = a$. For the flow external to a circular cylinder radius a, rotating with angular velocity Ω_a, a steady solution is only available if the circulation at large distances, Γ, is the same as at the cylinder so that $\Gamma = 2\pi a^2 \Omega_a$ with $v_\theta = a^2 \Omega_a / r$. Otherwise, as noted by Preston (1950), suction is required to maintain a steady flow with, again, $v_r = -m/r$ and

$$v_\theta = \frac{\Gamma}{2\pi r} + \left(a\Omega_a - \frac{\Gamma}{2\pi a} \right) \left(\frac{a}{r} \right)^{R-1}, \quad R \neq 2,$$

$$v_\theta = \frac{\Gamma}{2\pi r} + \left(a\Omega_a - \frac{\Gamma}{2\pi a} \right) \frac{a \ln r}{r \ln a}, \quad R = 2.$$

Accordingly, even when suction is applied, for $R \leq 2$ the only steady flow with finite circulation at infinity has $\Gamma = 2\pi a^2 \Omega_a$. But if the suction velocity exceeds $2\nu/a$ different values of the circulation may be maintained at infinity and the cylinder.

Berman (1958a) has considered axial flow driven by a constant pressure gradient along the annulus for $m \neq 0$ in the absence of rotation.

3.2 Non-circular pipe flow

For flow driven along a pipe of elliptic section, say $x^2/a^2 + y^2/b^2 = 1$, with a constant pressure gradient $-P$, Boussinesq (1868) found that

$$w(x, y) = \frac{Pa^2b^2}{2\mu(a^2 + b^2)} \left(1 - \frac{x^2}{a^2} - \frac{y^2}{b^2} \right),$$

with

$$Q = \frac{\pi Pa^3b^3}{4\mu(a^2 + b^2)}.$$

The eccentricity of the ellipse is $e = (1 - b^2/a^2)^{1/2}$ where a is the semi-major axis. The flow becomes Poiseuille flow in a circular pipe when $e = 0$, so that $a = b$, and plane Poiseuille flow in a channel as $a \to \infty$ with b fixed.

A wide variety of pipes with other cross-sections have been considered. For example, a sector of a circle with boundaries $r = 0, a$, $\theta = \pm\alpha$; a curvilinear rectangle bounded by two concentric circular arcs $r = a, b$ and two radii; the annulus bounded by two eccentric circles or two confocal ellipses; a lemniscate; a circle 'notched' by an arc of another circle; a limaçon; a cardioid. Berker (1963) gives a comprehensive bibliography and discusses some of these in detail.

3.3 Beltrami flows and their generalisation

As already noted in section 2.2, steady Beltrami flows for which $\mathbf{v} \wedge \boldsymbol{\omega} = \mathbf{0}$ can only exist in a viscous fluid when sustained by a non-conservative body force. The generalised Beltrami flows are required to satisfy equations (2.14) and (2.15). For axisymmetric flows we have $\mathbf{v} = (v_r, 0, v_z)$ and $\boldsymbol{\omega} = (0, \omega_\theta, 0)$. With $\boldsymbol{\omega} = \nabla \wedge \mathbf{v}$, and introducing the stream function ψ as in equation (1.24), we have

$$\frac{\partial}{\partial r}\left(\frac{1}{r}\frac{\partial \psi}{\partial r} \right) + \frac{1}{r}\frac{\partial^2 \psi}{\partial z^2} = -\omega_\theta, \tag{3.12}$$

whilst equations (2.14) and (2.15) require

$$\frac{\partial}{\partial r}\left(\frac{\omega_\theta}{r}\frac{\partial \psi}{\partial z} \right) - \frac{\partial}{\partial z}\left(\frac{\omega_\theta}{r}\frac{\partial \psi}{\partial r} \right) = 0, \tag{3.13}$$

and

$$\frac{\partial}{\partial r}\left\{ \frac{1}{r}\frac{\partial (r\omega_\theta)}{\partial r} \right\} + \frac{\partial^2 \omega_\theta}{\partial z^2} = 0. \tag{3.14}$$

Now, equation (3.13) has the solution $\omega_\theta = rf(\psi)$ showing that ω_θ/r is a constant along streamlines. However, Marris and Aswani (1977) have determined that the only possibility for these axisymmetric generalised Beltrami flows is that $f = $ constant, or $\omega_\theta = \alpha r$. That being so, equation (3.14) is satisfied identically, and the remaining equation (3.12) to be solved for ψ is

$$\frac{\partial^2 \psi}{\partial r^2} - \frac{1}{r}\frac{\partial \psi}{\partial r} + \frac{\partial^2 \psi}{\partial z^2} = -\alpha r^2. \tag{3.15}$$

A simple, but useful, set of solutions of equation (3.15) is

$$\psi = c_1 r^4 + c_2 r^2 z^2 + c_3 r^2 + c_4 r^2 z + c_5(r^6 - 12r^4 z^2 + 8r^2 z^4),$$

where $8c_1 + 2c_2 = -\alpha$, but the c_i are otherwise arbitrary. For various combinations consider the following.

(i) $c_1, c_4 \neq 0$. Without loss of generality we may take $\psi = ar^2(br^2 - z)$, Berker (1963), which corresponds to two opposing rotational streams divided by the paraboloidal stream surface $z = br^2$.

(ii) $c_2 \neq 0$. The solution $\psi = c_2 r^2 z^2$ is a special case of solutions obtained by Agrawal (1957), discussed in more detail in section 3.7. This particular solution may be interpreted as rotational flow against the plane boundary $z = 0$ at which the no-slip condition is satisfied.

(iii) $c_1, c_2, c_3 \neq 0$. Again without loss of generality, the solution may be taken as

$$\psi = ar^2 \left(\frac{r^2}{b^2} + \frac{z^2}{c^2} - 1 \right).$$

With $b \neq c$ this represents the ellipsoidal vortex of O'Brien (1961), whilst with $b = c$ we have Hill's (1894) spherical vortex with radius b. In the latter case the spherical vortex may be immersed in a stream, uniform at infinity, with the velocities, but not the stresses, continuous across the spherical interface, see Acheson (1990).

(iv) $c_1, c_3, c_5 \neq 0$.

In this case the solution, discovered by Wang (1990a), may be written as

$$\psi = r^2\{ar^2 + b + c(r^4 - 12r^2 z^2 + 8z^4)\}. \tag{3.16}$$

Solutions not dissimilar to Hill's spherical vortex, in the sense of a toroidal vortex bounded by a closed stream surface, can be constructed from (3.16). An example is given in figure 3.3 where a relatively elongated toroidal vortex is shown.

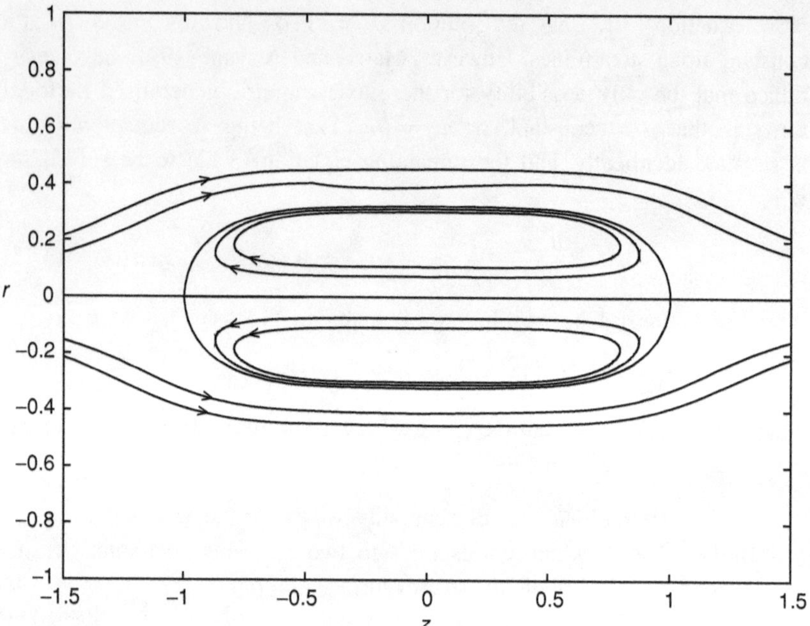

Figure 3.3 Wang's toroidal vortex, equation (3.16) with $a = 10$, $b = -1$, $c = \frac{1}{8}$. Commencing with the innermost the streamsurfaces have $\psi = -0.005$, $-0.01, 0, 0.1, 0.2$.

Additional solutions of the homogeneous form of equation (3.15) have been obtained by Berker (1963) as

$$\psi = r\left\{A_k J_1(kr) + B_k Y_1(kr)\right\}(C_k\, e^{kz} + D_k\, e^{-kz})$$

where J_1, Y_1 are Bessel functions of the first and second kinds respectively.

As a utilisation of this consider the flow represented by

$$\psi = \frac{Pa^4}{16\mu}\left\{2\left(\frac{r}{a}\right)^2 - \left(\frac{r}{a}\right)^4\right\} + \sum_{k=1}^{\infty} c_k\left(\frac{r}{a}\right) J_1(\lambda_k r/a)\, e^{\lambda_k z/a}. \qquad (3.17)$$

The first term of equation (3.17) may be recognised as the Poiseuille pipe flow considered in section 3.1, the second term, a distortion of this, will satisfy the no-slip condition at $r = a$ provided that the constants λ_k are the zeros of the Bessel function J_0. At $r = a$ the normal component of velocity is then given by

$$v_r|_{r=a} = \frac{1}{a^2}\sum_{k=1}^{\infty} c_k J_1(\lambda_k)\, e^{\lambda_k z/a}. \qquad (3.18)$$

The solution (3.17) was introduced by Terrill (1982) who discusses various situations in which (3.18) represents the transpiration velocity across the porous pipe $r = a$.

Additional solutions which include the effects of swirl in generalised Beltrami flows have been discussed by Weinbaum and O'Brien (1967).

3.4 Stagnation-point flows

3.4.1 The classical Homann (1936) solution

In section 2.3 we discussed the flow in the immediate vicinity of the stagnation line on a rigid stationary cylinder. The axisymmetric analogue of that arises, for example, at the front stagnation point of a sphere placed in a uniform stream. Homann (1936) modelled this by the flow towards an infinite, rigid, stationary flat plate. This is, of course, a special case of the three-dimensional stagnation-point flow of section 2.5.4, but we include it here for completeness. In our formulation we allow the plate to slide in its own plane with constant velocity, and also allow for transpiration across it when porous; both of these features have been examined by Libby (1974, 1976) in the more general context of a three-dimensional stagnation point, whilst Wang (1973) has considered axisymmetric flow against a sliding plane.

The boundary is taken as $z = 0$, and with no natural length scale in the problem a self-similar solution with velocity components

$$v_r = krf'(\eta) + u_w \cos\theta g(\eta), \quad v_\theta = -u_w \sin\theta g(\eta),$$

$$v_z = -2(kv)^{1/2} f(\eta), \qquad \eta = \left(\frac{k}{v}\right)^{1/2} z, \qquad (3.19)$$

and pressure

$$\frac{p - p_0}{\rho} = -\frac{1}{2}k^2 r^2 - 2kv(f^2 + f'), \qquad (3.20)$$

leads, using equations (1.19), (1.20), to the following equations for f and g,

$$f''' + 2ff'' - f'^2 + 1 = 0,$$
$$g'' + 2fg' - f'g = 0,$$

with boundary conditions

$$f(0) = \lambda, \quad f'(0) = 0, \quad f'(\infty) = 1, \quad g(0) = 1, \quad g(\infty) = 0.$$

With $\lambda = u_w = 0$ we recover the classical stagnation-point flow of Homann (1936). With $u_w \neq 0$ we have, without any loss of generality, the boundary

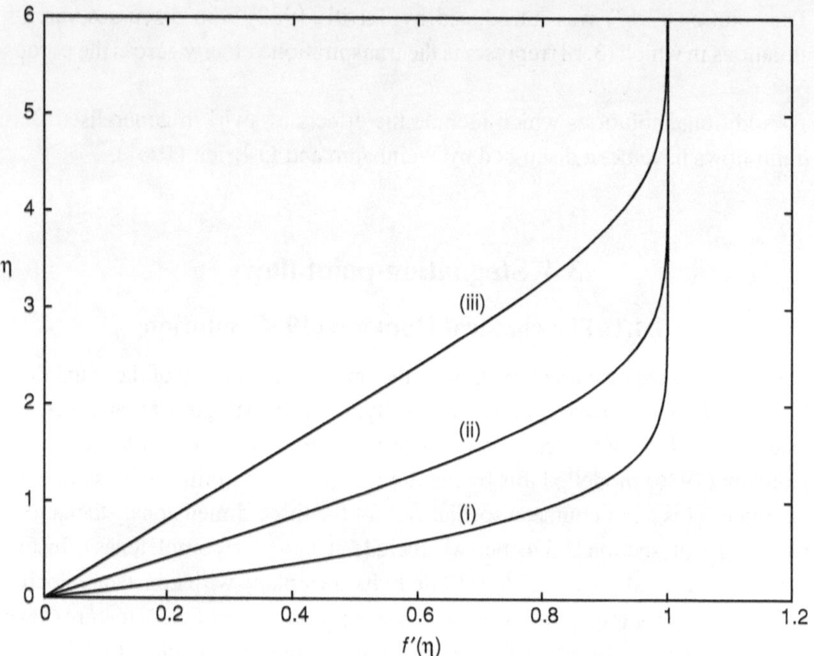

Figure 3.4 The radial velocity function $f'(\eta)$ at a stagnation point. (i) The case $\lambda = 0$ corresponds to the classical Homann flow, (ii) $\lambda = -1$, (iii) $\lambda = -2$.

sliding in its own plane with constant speed in the direction $\theta = 0$; this corresponds to the case addressed by Libby (1974) at a three-dimensional stagnation point, and by Wang (1973). For $\lambda \neq 0$ the boundary is assumed to be porous with transpiration across it, again a case considered by Libby (1976) (in fact, for a compressible fluid) at a three-dimensional stagnation point. For $\lambda < 0$ we have injection, perhaps the more interesting case, and in figure 3.4 we show profiles $f'(\eta)$ for various values of λ. As the injection rate increases the position η_0, where $f(\eta_0) = 0$, increases, with viscous effects increasingly unimportant for $\eta < \eta_0$. Corresponding profiles $g(\eta)$ are shown in figure 3.5. For $\eta < \eta_0$ there is again, relatively speaking, a lack of structure in the solution with a fairly rapid change taking place in the neighbourhood of $\eta = \eta_0$.

Davey (1963) has considered the solution at a three-dimensional stagnation point when the oncoming flow is rotational, and when the boundary itself may rotate. Here we present only the axisymmetric stagnation point, at a stationary

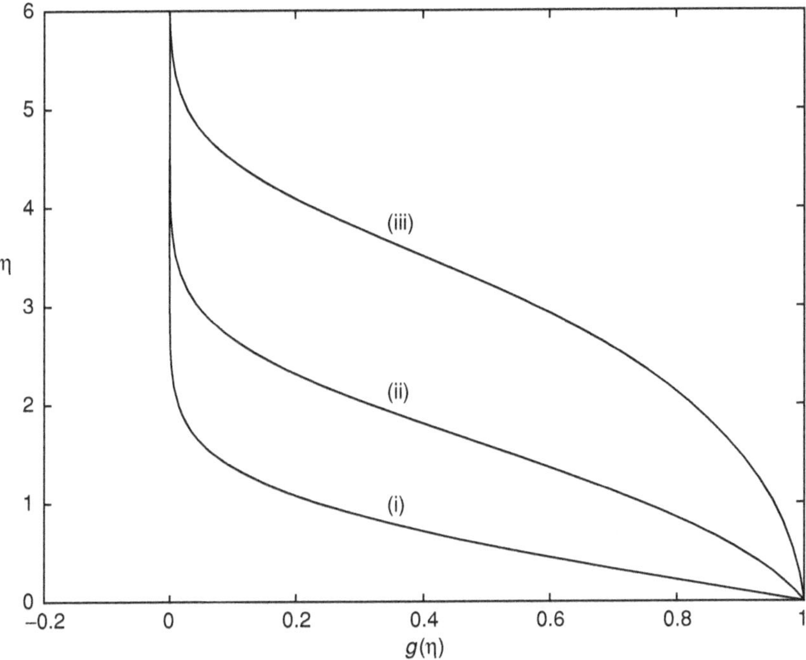

Figure 3.5 The velocity function $g(\eta)$ corresponding to boundary translation with speed u_w; see equation (3.19). (i) $\lambda = 0$, (ii) $\lambda = -1$, (iii) $\lambda = -2$.

plane boundary, with $\boldsymbol{\omega} = 2\Omega\mathbf{k}$ in the oncoming flow. Equations (3.19), (3.20) are replaced by

$$v_r = krf'(\eta), \quad v_\theta = \Omega r g(\eta), \quad v_z = -2(k\nu)^{1/2}f(\eta), \quad \text{with again } \eta = \left(\frac{k}{\nu}\right)^{1/2} z,$$

and

$$\frac{p - p_0}{\rho} = \frac{1}{2}k^2 r^2\left\{\left(\frac{\Omega}{k}\right)^2 - 1\right\} - 2\nu k(f' + f^2);$$

in addition a non-conservative body force is required to sustain this motion with $\mathbf{F} = 2\Omega k r\,\hat{\boldsymbol{\theta}}$. Ordinary differential equations for f and g are again derived from equations (1.19) and (1.20) as

$$f''' + 2ff'' - f'^2 + (\Omega/k)^2 g^2 + 1 - (\Omega/k)^2 = 0, \tag{3.21}$$

$$g'' + 2fg' - 2f'g + 2 = 0, \tag{3.22}$$

with boundary conditions

$$f(0) = f'(0) = g(0) = 0, \quad f'(\infty) = 1, \quad g(\infty) = 1. \tag{3.23}$$

Solutions of equations (3.21), (3.22) subject to (3.23) are shown in figure 3.6.

A practical application of the stagnation-point solution has been identified by Hinch and Lemaître (1994) to the problem of disks floating above an air table through which air is blown. Suppose two infinite planes are at a distance h apart. The upper is impermeable, the lower porous through which fluid is injected normally with uniform speed W. Then, if equations (3.19), (3.20) are replaced by

$$v_r = -\frac{Wr}{2h} f'(\eta), \quad v_z = Wf(\eta),$$

$$\frac{p - p_0}{\rho} = -\frac{1}{2}\frac{W^2}{h^2} r^2 \beta + \frac{Wv}{h}\left(f' - \frac{1}{2} R f^2 \right),$$

where $\eta = z/h$, $R = Wh/v$, then equations (1.19) and (1.21) will be satisfied provided that

$$\frac{1}{R} f''' - ff'' + \frac{1}{2} f'^2 - 2\beta = 0, \tag{3.24}$$

which is to be solved subject to

$$f(0) = 1, \quad f'(0) = 0, \quad f(1) = f'(1) = 0.$$

Four boundary conditions for the third-order equation (3.24) reflect the fact that β is unknown *a priori*. Hinch and Lemaître present solutions across the whole range of Reynolds number R. Cox (2002) has extended the above to non-axisymmetric flow using the formulation of Howarth (1951) and Davey (1961). He concludes that for given W and h the non-axisymmetric flow, if achievable, can support a greater weight than the axisymmetric flow.

3.4.2 Stagnation on a circular cylinder

We have discussed the classical symmetric stagnation-point flows of Hiemenz (1911) and Homann (1936), and the asymmetric flows that link them detailed by Howarth (1951) and Davey (1961). A somewhat different stagnation flow was discovered by Wang (1974), namely that on a circle of radius a, circumscribing a circular cylinder of the same radius whose generators lie in the z-direction. For an inviscid fluid it is easily verified that this stagnation flow has

$$v_r = -k\left(r - \frac{a^2}{r} \right), \quad v_z = 2kz, \quad \frac{p - p_0}{\rho} = -2k^2 z^2 - \frac{1}{2} k^2 \left(r - \frac{a^2}{r} \right)^2. \tag{3.25}$$

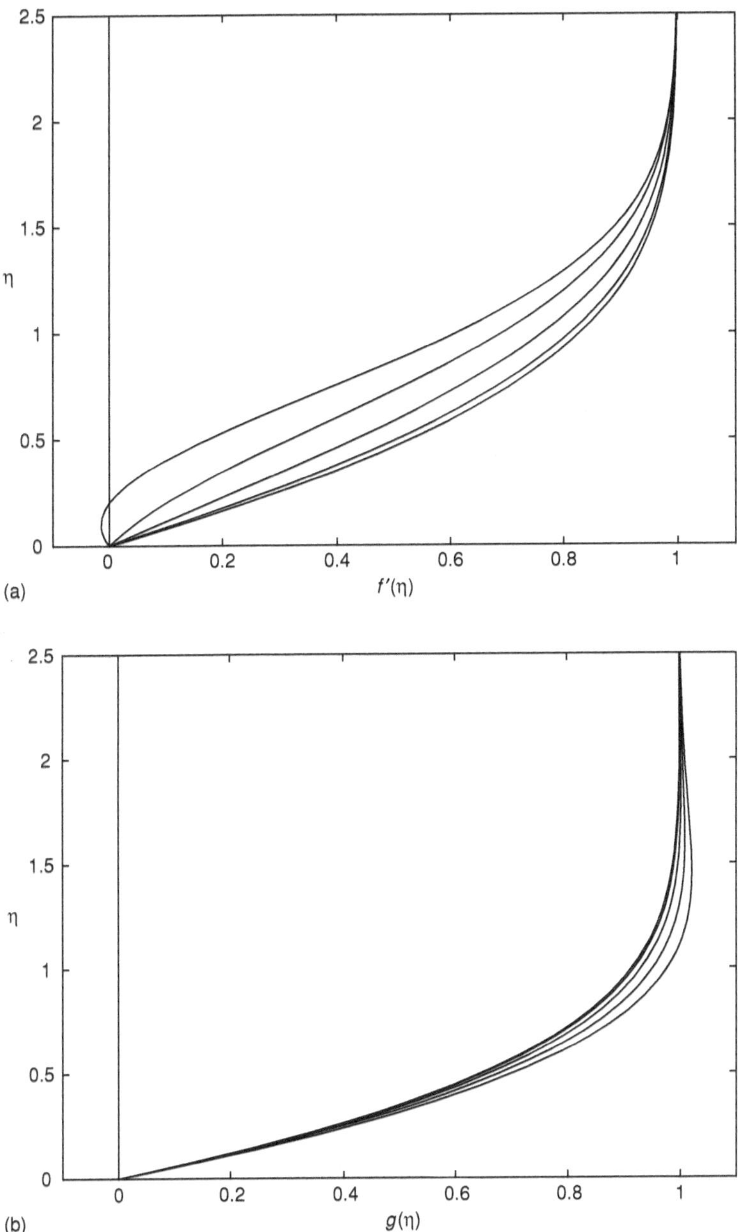

Figure 3.6 (a) The radial velocity function $f'(\eta)$ at a rotational stagnation point. Successive profiles from the lowest, which corresponds to the classical Homann flow, have $\Omega/k = 0.0(0.5)2.0$. (b) The azimuthal velocity function $g(\eta)$ corresponding to (a).

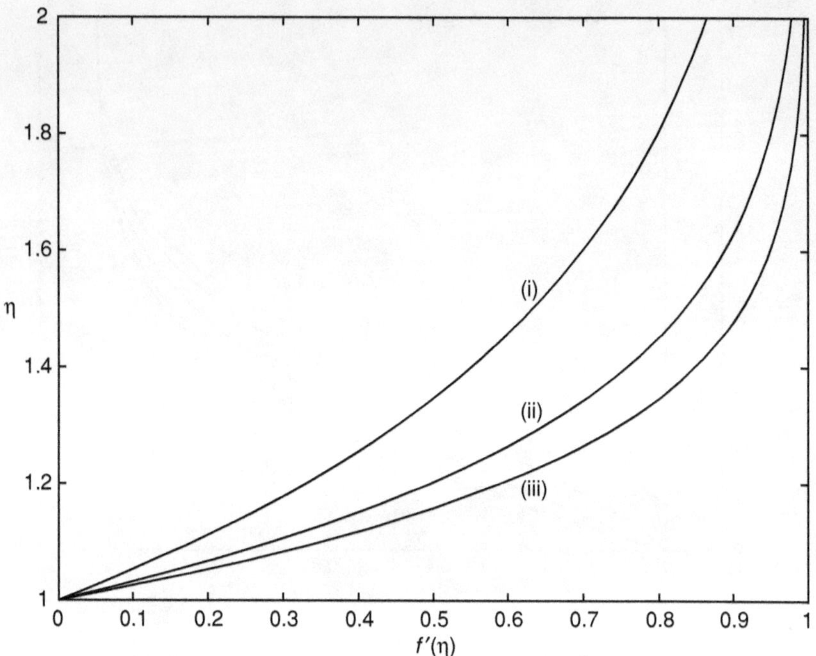

Figure 3.7 Profiles of the axial velocity function $f'(\eta)$ for stagnation flow on a circular cylinder. (i) $R = 2$, (ii) $R = 6$, (iii) $R = 10$.

For a viscous fluid we may write the Stokes stream function for this rotationally symmetric flow as

$$\psi = ka^2 z f(\eta) \quad \text{where} \quad \eta = (r/a)^2, \tag{3.26}$$

so that, from equation (1.24), we have the velocity components

$$v_r = -ka\eta^{-1/2} f(\eta) \quad \text{and} \quad v_z = 2kzf'(\eta). \tag{3.27}$$

Substituting into equations (1.19) and (1.20) gives, as the equation for f,

$$\eta f''' + f'' + R(ff'' - f'^2 + c) = 0, \tag{3.28}$$

where $R = ka^2/2\nu$ is a Reynolds number, and for the pressure

$$\frac{p - p_0}{\rho} = -\frac{1}{2}k^2a^2\eta^{-1}f^2(\eta) - 2\nu kf'(\eta) - 2k^2z^2.$$

The boundary conditions for (3.28) require $f(1) = f'(1) = 0$, and for consistency with equation (3.25), $c = f'(\infty) = 1$. Equation (3.28) has been integrated numerically by Wang, and we present some such solutions in figure 3.7. We may note that as R increases a boundary-layer structure emerges. Indeed Wang has shown that for $R \gg 1$ equation (3.28) may be transformed to the two-dimensional Hiemenz equation with relative error $O(R^{-1/2})$.

Burde (1994) has *inter alia*, also addressed this problem and finds an exact solution, for the case $R = 2$, as

$$f(\eta) = \tfrac{1}{2}\{2\eta + e^{2(1-\eta)} - 3\},$$

which is included in figure 3.7.

Variations of this basic flow have been considered by several authors, Wang himself has considered the analogue of Crane's (1970) stretching plate problem. In that case the appropriate solution is still as in equations (3.26) and (3.27) with $f(\eta)$ satisfying (3.28), but now the boundary conditions are

$$f(1) = 0, \quad f'(1) = 1, \quad f'(\infty) = c = 0. \tag{3.29}$$

In obtaining numerical solutions, Wang draws attention to the difficulty in this case that solutions decay algebraically, rather than exponentially, as $\eta \to \infty$. Again Burde (1989) has found an exact solution in one particular case. In addition to the stretching cylinder, Burde introduces a uniform flow with speed U_∞ parallel to the generators of the cylinder. In that case equation (3.26) is replaced by

$$\psi = ka^2 z f(\eta) + U_\infty a^2 \int_0^\eta g(s)\,\mathrm{d}s.$$

The equation satisfied by $f(\eta)$ is again equation (3.28), with boundary conditions (3.29), whilst $g(\eta)$ satisfies

$$\eta g'' + g' + R(fg' - f'g) = 0, \tag{3.30}$$

with

$$g(1) = 0, \quad g(\infty) = \tfrac{1}{2}. \tag{3.31}$$

Burde notes that for the special case $R = 4/3$ there is an exact solution of each of equations (3.28) and (3.30), subject to (3.29) and (3.31) respectively, as

$$f(\eta) = 3\left(\frac{\eta - 1}{2\eta + 1}\right),$$

$$g(\eta) = \frac{4\eta^3 - 6\eta^2 - 4\eta + 1}{(2\eta + 1)^3} + \frac{3\ln\{(2\eta + 1)/3\} + 5/3}{(2\eta + 1)^2}.$$

Axial velocity profiles for this special case are shown in figure 3.8. Mention may also be made of the work of Gorla (1978) who, in the original problem of Wang (1974), allows the cylinder to slide parallel to its generators with uniform speed.

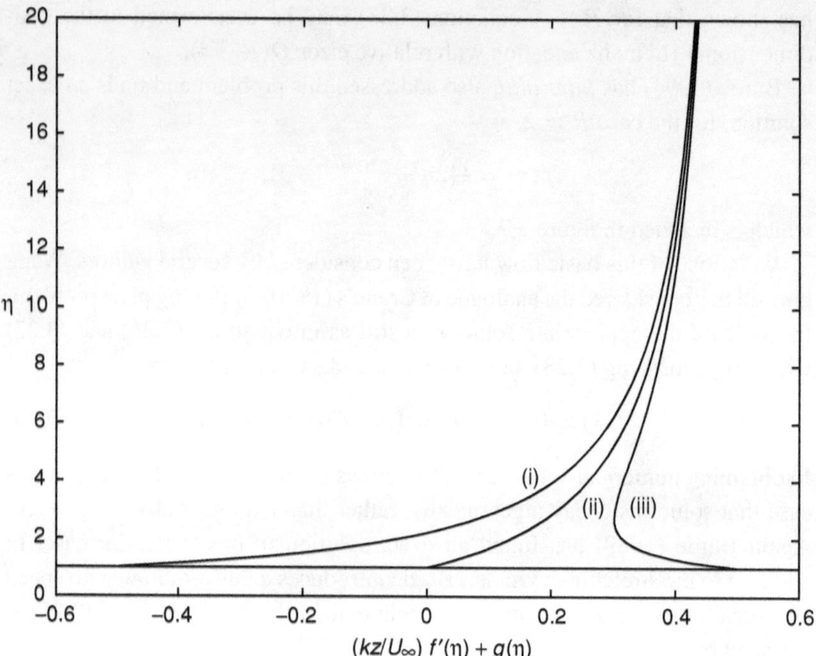

Figure 3.8 The axial velocity profile $v_z/2U_\infty = (kz/U_\infty)f'(\eta) + g(\eta)$ at a stretching cylinder in a uniform axial flow for $R = \frac{4}{3}$. (i) $kz/U_\infty = -\frac{1}{2}$, (ii) $kz/U_\infty = 0$, (iii) $kz/U_\infty = \frac{1}{2}$.

An analogue of the oblique stagnation-point flow considered in section 2.3.2 has been discussed by Weidman and Putkaradze (2003, 2005). In the far field the velocity components (3.25) are replaced by

$$v_r = -k\left(r - \frac{a^2}{r}\right), \quad v_z = 2kz + \gamma\left\{\left(\frac{r}{a}\right)^2 - 1\right\}.$$

The far-field vorticity is $-2\gamma r\hat{\theta}$ and an additional term $4\gamma vz/a^2$ is required in the pressure field. Unlike the corresponding two-dimensional case, it should be noted that viscous effects are not negligible in the far field. Replacing (3.25) by

$$v_r = -ka\eta^{-1/2}f(\eta), \quad v_z = 2kzf'(\eta) + \gamma g(\eta),$$

and the corresponding pressure field by

$$\frac{p - p_0}{\rho} = -\frac{1}{2}k^2a^2\eta^{-1}f^2(\eta) - 2vkf'(\eta) - 2k^2z^2 + \frac{4\gamma v}{a^2}z,$$

where again $\eta = (r/a)^2$, then the equation for $f(\eta)$ is unchanged whilst, from (1.11), $g(\eta)$ satisfies

$$\eta g'' + g' + R(fg' - f'g) = 1, \quad \text{with} \quad g(0) = 0, \quad g'(\infty) = 1.$$

Solutions have been obtained by Weidman and Putkaradze (2005) for a range of values of R.

In a further development Cunning, Davis and Weidman (1998) elaborate upon the solution of Wang (1974) by both introducing uniform transpiration V_0 at the cylinder surface, and allowing it to rotate with constant angular velocity Ω. The solution is of the form (3.26) where $f(\eta)$ satisfies (3.28), supplemented by the azimuthal velocity

$$v_\theta = \Omega a \eta^{-1/2} h(\eta)$$

where $h(\eta)$ satisfies, from equation (1.20),

$$\eta h'' + R f h' = 0, \tag{3.32}$$

and the pressure is now given by

$$\frac{p - p_0}{\rho} = -\frac{k^2 a^2 f(\eta)}{2\eta} - 2\nu k f'(\eta) - 2k^2 z^2 + \frac{\Omega^2 a^2}{2} \int_1^\eta \frac{g^2(s)}{s^2} ds.$$

The boundary conditions for equations (3.28) and (3.32) now require

$$f(1) = -S, \quad f'(1) = 0, \quad f'(\infty) = 1, \quad h(1) = 1, \quad h(\infty) = 0, \tag{3.33}$$

where the transpiration parameter $S = V_0/ka$. For the case in which there is no rotation, $\Omega \equiv 0$, Burde (1994) has again presented exact solutions of equation (3.28) for $R = 2/(1 + S)$ as

$$f(\eta) = \eta - \tfrac{3}{2}(1 + S) + \tfrac{1}{2}(1 + S) e^{2(1-\eta)/(1+S)}.$$

Although this solution encompasses a wide range of values of R and S, the two parameters cannot be prescribed independently. Cunning *et al.* (1998) have solved equation (3.28) numerically and, simultaneously, equation (3.32), subject to the conditions (3.33) for a wide range of values of the parameters R and S. Of particular interest are the solutions for strong blowing, $S \gg 1$ with R fixed. The flow structure that develops is one in which the axial profile $f'(\eta)$ thickens without change of character. However, the swirling velocity component, within the surface of zero radial velocity which itself increases in radius linearly with S, exhibits the character of a potential vortex. At the surface of zero radial velocity this is destroyed and the swirl decays rapidly to zero across a viscous layer of thickness $O\{(RS)^{-1/2}\}$.

3.4.3 Flow inside a porous or stretching tube

For the case of a porous, or stretching, tube it is not unnatural to consider the flow induced within it, that is, the analogue of the flow between plates across which there is suction or injection, or which stretch as considered in section 2.4.1. In the planar case the Hiemenz transformation was seen to be appropriate for fully developed flow and it is Wang's transformation, as in equation (3.26), that is the starting point for our discussion of flow within the tube.

Consider first the case of a porous tube across whose boundary fluid is either sucked or injected. The ordinary differential equation for the fully developed flow was initially addressed numerically by Berman (1958b) and White (1962). The solutions indicated that for wall suction solutions were not always possible, suggesting that in those cases fully developed flow is not possible, and that where solutions were obtained a dual was always available. However, the most comprehensive and penetrating account which reveals the rich and complex structure of the flow has been given by Terrill and Thomas (1969). Following Terrill and Thomas we write the analogue of equations (2.52) and (2.53) for this axisymmetric flow as

$$\psi = a^2(U_0/2 - V\bar{z})f(\eta), \quad \eta = (r/a)^2, \tag{3.34}$$

where $\bar{z} = z/a$, V is the velocity of suction at the tube wall $r = a$ and U_0 is an arbitrary constant velocity at $\bar{z} = 0$, though it may be remarked that Terrill and Thomas also explore the possibility of a non-uniform axial velocity at $\bar{z} = 0$. With $\mu V/a$ as a scale for the pressure we have as the analogue of equation (2.53)

$$\bar{p} = \frac{p}{\rho V^2 R^{-1}} = 2f' - \frac{1}{2}R\eta^{-1}f^2 - 4\lambda\bar{z}^2 + 4\lambda U_0 V^{-1}\bar{z},$$

where $R = Va/\nu$ is the Reynolds number, positive when fluid is sucked from the tube, negative when fluid is injected into it, and λ is not determined *a priori*. The equation satisfied by $f(\eta)$ is, analogous to (2.54),

$$\eta f''' + f'' + \tfrac{1}{2}R(f'^2 - ff'') - \lambda = 0, \tag{3.35}$$

with boundary conditions

$$f(0) = 0, \quad \lim_{\eta \to 0} \eta^{1/2} f''(\eta) = 0, \quad f(1) = 1, \quad f'(1) = 0. \tag{3.36}$$

As in the case of plane boundaries four boundary conditions are required, since λ also has to be determined. The principal results of Terrill and Thomas are that for $2.3 < R < 9.1$ there cannot be a fully developed flow in the tube, as in equation (3.34), that for $R < 2.3$ and for $9.1 < R < 20.6$ there are two

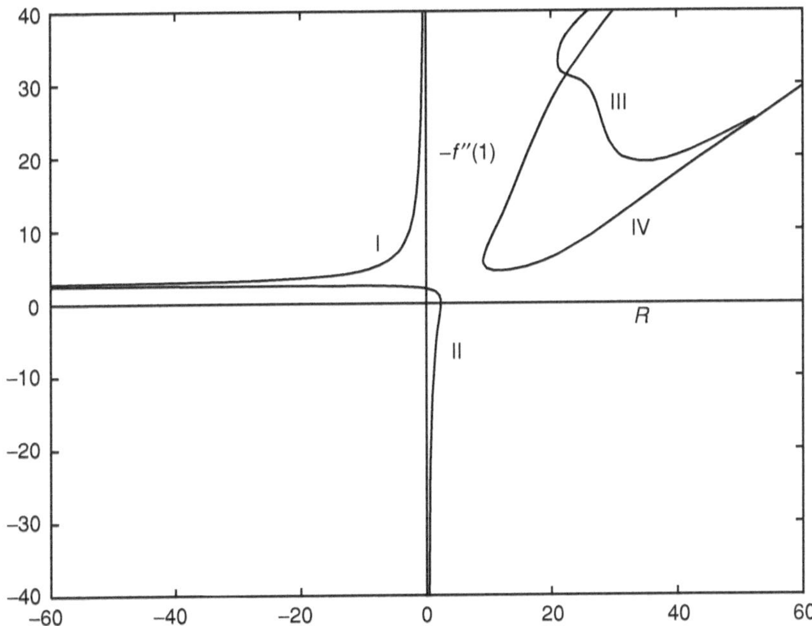

Figure 3.9 A map of the available solutions for flow within a porous tube in the $[-f''(1), R]$-plane. The solution groups I, II, III and IV are described in the text and illustrated below.

solutions. Skalak and Wang (1977) have extended the results of Terrill and Thomas and shown that for $R > 20.6$ there are four solutions. These results are shown in figure 3.9, where the variation of $-f''(1)$ is shown with Reynolds number R.

There are, essentially, four groups of solution corresponding to the four solution branches, and we consider each in turn with designations that differ from those of Terrill and Thomas. The solutions in Group I cover the whole range of injection Reynolds numbers. These solutions are characterised by a region of reversed flow in the central part of the tube. For large injection, the velocity profiles are unexceptional, but as $R \to 0-$ both the centre-line velocity $f'(0)$ and the wall shear stress $f''(1)$ tend to minus infinity. Examples of these Group I solutions are shown in figure 3.10.

On the lower branch of injection solutions in figure 3.9, designated Group II, $f'(\eta) > 0$ for all η with little variation in the profiles and with $f''(1)$ increasing. The further development on that solution branch shows R at first increasing into the suction regime, with a region of reversed flow developing in the wall region, and then decreasing. As $R \to 0 +$ both the centre-line velocity and wall

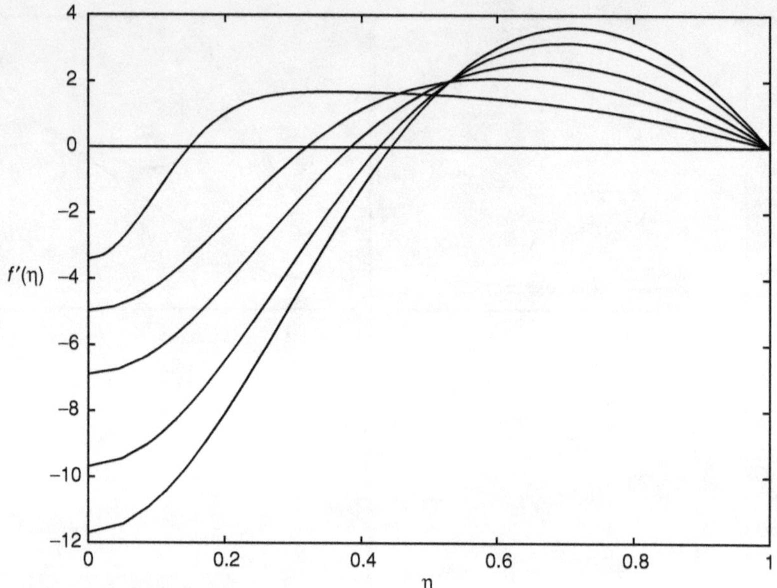

Figure 3.10 Velocity profiles $f'(\eta)$ corresponding to Group I in figure 3.9 with, from the lowest at $\eta = 0$: $R = -2.01, -2.61, -4.44, -8.83, -57.83$.

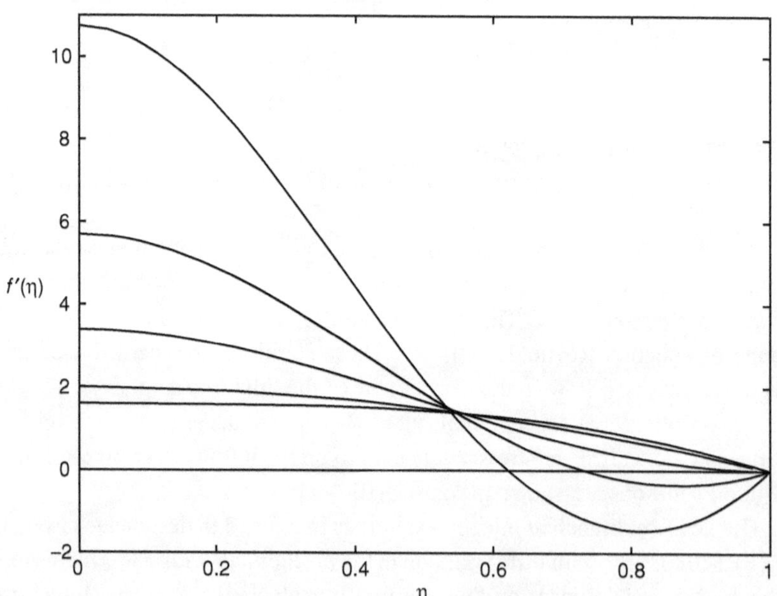

Figure 3.11 Velocity profiles $f'(\eta)$ corresponding to Group II in figure 3.9 with, from the lowest at $\eta = 0$: $R = -52.2, -0.04, 2.30, 1.93, 1.27$.

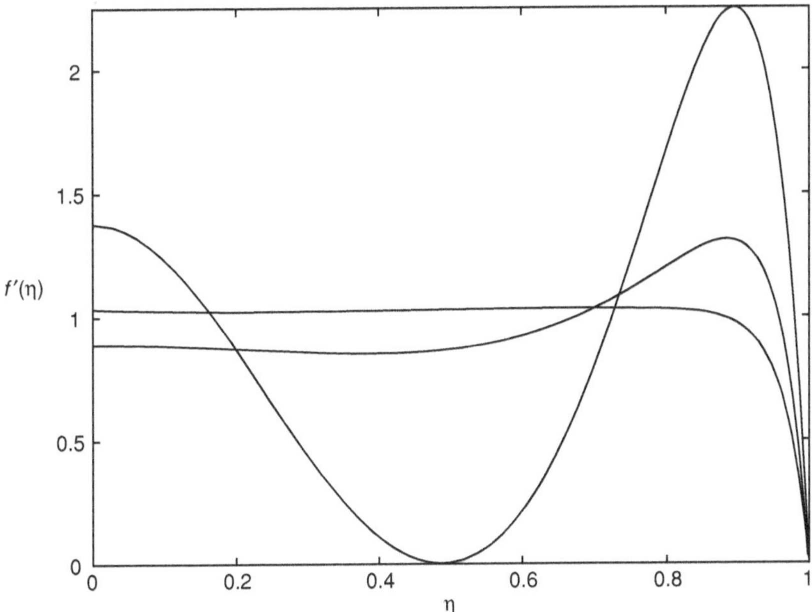

Figure 3.12 Velocity profiles $f'(\eta)$ corresponding to Group III in figure 3.9 with, from the lowest at $\eta = 0$: $R = 30.61, 60.0, 23.75$.

shear stress tend to infinity; typical profiles are shown in figure 3.11. Terrill and Thomas show that the suction and injection profiles of Group II and Group I, as $R \to 0 \pm$ respectively, are the images of one another, reflected in the η-axis.

As already noted there is then a range of suction Reynolds numbers for which no solutions of equation (3.35) exist. Consider next the upper branch of solutions for $R > 0$, Group III. For $R \gg 1$ there is a core of uniform flow, with $f'(\eta) \approx 1$, flanked by a boundary layer of thickness $O(R^{-1})$. As R decreases the profile develops two points of inflection with local maxima close to the boundary and at the centre-line, with $f'(\eta) \lesssim 1$ respectively, and a minimum between them; further decrease in R shows $f'(0)$ increasing to values greater than unity with the maximum close to $\eta = 1$ increasing and the minimum between them dipping below zero. Examples are shown in figure 3.12.

Finally there are Group IV solutions corresponding to the lower branch of suction solutions in figure 3.9. For $R \gg 1$ these are virtually indistinguishable from those of Group III. As R decreases the centre-line velocity at first increases and then decreases until for $R \approx 10$ there is reversed flow on the centre-line. Meanwhile close to the boundary the velocity increases. This development continues beyond the minimum value $R = 9.1$ and as R again increases on the

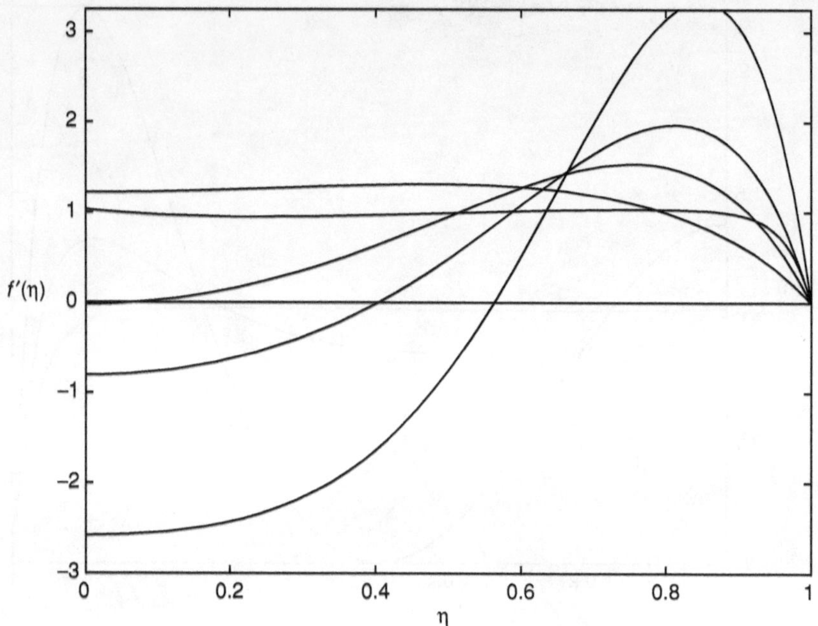

Figure 3.13 Velocity profiles $f'(\eta)$ corresponding to Group IV in figure 3.9 with, from the lowest at $\eta = 0$: $R = 21.6, 13.06, 10.04, 60.0, 13.83$.

upper part of this solution branch, until as $R \to 21.2$ the solution develops a positive maximum close to the centre-line. At $R = 21.2$ the solution is almost identical with that of Group III. A closer examination reveals, however, that as $R \to 21.2+$, $f''(0) \to \infty$ whilst as $R \to 21.1-$, $f''(0)$ remains finite. Velocity profiles for Group IV solutions are shown in figure 3.13.

As has been remarked the suction solutions of Groups III and IV are virtually indistinguishable for $R \gg 1$. Terrill and Thomas indicate that the two solutions differ by exponentially small terms. This difference is examined in more detail in a subsequent paper by Terrill (1973), who has also considered flow through a porous annulus (Terrill (1966)), extending the work of Berman (1958a). This annular flow problem has been extended by Marques, Sanchez and Weidman (1998) to include rotation, translation, stretching and twisting of the cylinders.

The absence of any solutions of equation (3.35), representing fully developed flow, in the range of Reynolds numbers $2.3 < R < 9.1$ has not been explained. Prager (1964) introduced a swirling component of velocity into the tube such that

$$v_\theta = (U_0 - 2V\bar{z})\eta^{-1/2}g(\eta),$$

with v_r, v_z again determined from the stream function (3.34). The equation satisfied by g is

$$\eta g'' + \tfrac{1}{2}R(f'g - fg') = 0$$

with $g(0) = 0$, $g(1) = 1$. This equation and that governing the axial flow, which now includes centrifugal effects, have been solved numerically by Terrill and Thomas (1973). Again there is a rich solution structure upon which we comment as follows. For injection, $R < 0$, there are two solutions, both with $f''(1) > 0$ and for both of which $f''(1) \to 0$ as $R \to 0$. For $R \lesssim -6$ these proved extremely difficult to determine and were not pursued. For suction, $R > 0$, there are dual solutions, with $f''(1) < 0$, in the range $0 < R \le 9.7$ with only one solution for $R > 9.7$ which, as $R \to \infty$, becomes essentially identical with those of Groups III and IV in the absence of swirl. The differences are again attributed to exponentially small terms. The point of interest is, then, that the introduction of swirl into the tube leads to solutions of the fully developed flow equations for all values of $R > 0$. Though this sheds no light on why there should be a range of values of R in which no solutions are available in the absence of swirl.

As in the two-dimensional case of flow between porous or stretching plates there are some similarities between the flow driven by injection, discussed above, and the stretching tube flow considered by Brady and Acrivos (1981). With $\psi = U_0 a^2 \bar{z} f(\eta)$, and $\mu U_0/a$ again as a scale for the pressure, so that

$$\bar{p} = 4\lambda \bar{z}^2 - \left(2f' + \tfrac{1}{2}R\eta^{-1}f^2\right),$$

where $R = U_0 a/\nu$ and λ is undetermined, we now have as the equation for f,

$$\eta f''' + f'' + \tfrac{1}{2}R(ff'' - f'^2) - \lambda = 0, \tag{3.37}$$

together with

$$f(0) = 0, \quad \lim_{\eta \to 0} \eta^{1/2} f''(\eta) = 0, \quad f(1) = 0, \quad f'(1) = 1. \tag{3.38}$$

It is found numerically that equation (3.37), subject to (3.38), has no solutions in the range $10.25 < R < 147$. For $R < 10.25$ there are dual solutions. On one branch, as R increases from zero, $f'(0) < 0$ and decreases with $f'(\eta)$ increasing monotonically to $f'(1) = 1$ up to $R = 10.25$. As R decreases on the other branch from 10.25, $f'(0)$ decreases dramatically and is $O(R^{-1})$ as $R \to 0$. Simultaneously $f'(\eta)$ develops a maximum, close to the moving boundary, which itself increases dramatically as required by continuity.

These solutions for $R < 10.25$ are designated Group I. For $R > 147$ multiple solutions appear. Group II solutions are characterised for $R \gg 1$ by a boundary layer, thickness $O(R^{-1/2})$, at the moving boundary surrounding an inviscid core of uniform flow with $0 > f'(\eta) = O(R^{-1/2})$. As R decreases $f'(0)$ decreases until $R \approx 147$ whereupon R again increases, as does $f'(0)$, until $R = 186$ when $f'(0) = 0$. The Group III solutions are a continuation, with $f'(0)$ increasing with R, on this solution branch. On another solution branch there are again dual solutions designated Group IV. On one part of this branch when $R \gg 1$ the flow again consists of an inviscid core of zero vorticity surrounded by a conventional boundary layer. As R decreases on this solution branch the profile develops in a surprising manner such that for $R < 900$ there is a region of flow midway between the axis and the wall in which $f'(\eta) > 0$ flanked by regions with $f'(\eta) < 0$. This part of the solution branch ends at $R \approx 770$, and as R again increases on the other part the development described above continues with three different regions of flow in the core, in each of which the velocity increases with R, flanked by a boundary layer at the moving wall. In all cases, for $R > 147$, the values of $f''(1)$ are virtually indistinguishable.

Following Terrill and Thomas (1973), Brady and Acrivos introduce swirl into the tube and demonstrate the existence of dual solutions for all $R > 0$. However, as with the porous tube, this sheds no further light on the 'gap' range of Reynolds numbers when swirl is absent.

The flow outside a stretching tube has been considered by Wang (1988).

3.5 Rotating-disk flows

When a finite disk rotates in its own plane about an axis through its centre, in a fluid otherwise at rest, fluid is centrifuged radially and thrown off the perimeter in a jet-like manner, a phenomenon that is easily realised experimentally. As a model for this an infinite rotating disk was first studied by von Kármán (1921), and by Cochran (1934). Subsequently, extensions of this basic flow were proposed by Hannah (1947) who considered forced flow against the rotating plane, Batchelor (1951) who allowed solid-body rotation of the fluid in the main body of fluid beyond the plane and Stuart (1954) who considered the effect of uniform suction at it, when assumed porous. In a related situation a stationary plane is placed in a fluid otherwise in solid-body rotation; this was first discussed by Bödewadt (1940); whilst Batchelor (1951) studied the flow bounded by two disks, both of which were allowed to rotate. It is convenient to consider the one- and two-disk situations separately.

3.5.1 The one-disk problem

We concentrate on the axisymmetric flow first proposed by von Kármán, but we also consider a non-axisymmetric flow discovered by Hewitt, Duck and Foster (1999).

As with the stagnation-point flows considered in section 3.4 there is no natural length scale in this problem and a self-similar solution is available. Again utilising cylindrical polar co-ordinates with the rotating plane, angular velocity Ω, coinciding with $z = 0$, and the axis of rotation taken as $r = 0$, the velocity components may be written as

$$v_r = r\Omega f'(\eta), \quad v_\theta = r\Omega g(\eta), \quad v_z = -2(\nu\Omega)^{1/2} f(\eta), \quad \eta = \left(\frac{\Omega}{\nu}\right)^{1/2} z,$$
(3.39)

and the pressure as

$$\frac{p - p_0}{\rho} = \frac{1}{2} r^2 \Omega^2 \lambda - 2\nu\Omega(f^2 + f').$$
(3.40)

Equations (1.19), (1.20) then yield, as equations for f and g,

$$f''' + 2ff'' - f'^2 + g^2 = \lambda,$$
(3.41)

$$g'' + 2(fg' - f'g) = 0,$$
(3.42)

with boundary conditions

$$f(0) = \lambda_1, \quad f'(0) = 0, \quad g(0) = 1, \quad f'(\infty) = \lambda_2, \quad g(\infty) = \lambda_3. \quad (3.43)$$

In these equations we recover the classical von Kármán flow by setting $\lambda = \lambda_i = 0$ $(i = 1, 2, 3)$. Von Kármán himself derived an approximate method of solution, whilst the first adequate numerical solution was presented by Cochran using appropriate series expansions for large and small values of η. This classical solution is shown in figure 3.14; it clearly reveals the radial flow associated with centrifugal effects.

Stuart (1954) extended Cochran's numerical work to include the effect of suction at a porous disk with $\lambda_1 = 1$; this solution is also shown in figure 3.14 where we see the fluid is literally 'sucked' towards the disk when compared with the classical case. Stuart also presented a series solution in descending powers of λ_1.

With $\lambda = -(k/\Omega)^2$, $\lambda_2 = k/\Omega$, $\lambda_1 = \lambda_3 = 0$, the problem is formulated as by Hannah for the case of flow towards the rotating disk such that at large distances from it $v_r = kr$, $v_\theta = 0$, $v_z = -2kz$, as in the classical stagnation-point flow. Hannah calculated a few solutions for this case, whilst Tifford and Chu (1952) extended the range. One such solution is included in figure 3.15, showing a 'squeezing' of the flow towards the rotating plane. An extension of this to a case in which the onset flow is of saddle-point type, rather than

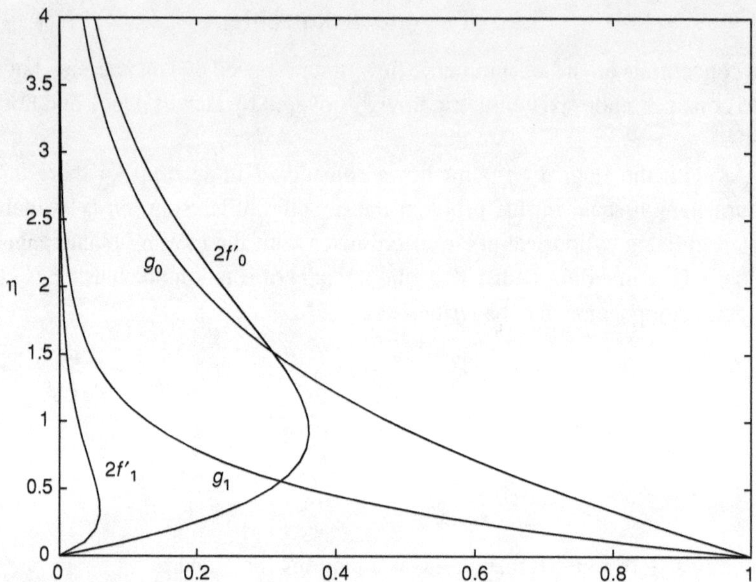

Figure 3.14 Rotating-disk flows. The radial and azimuthal velocity profiles $f_i'(\eta)$, $g_i(\eta)$ are shown for the classical von Kármán flow, $i = 0$, and for a case of suction at the boundary, $i = 1$, as considered by Stuart for $\lambda_1 = 1$.

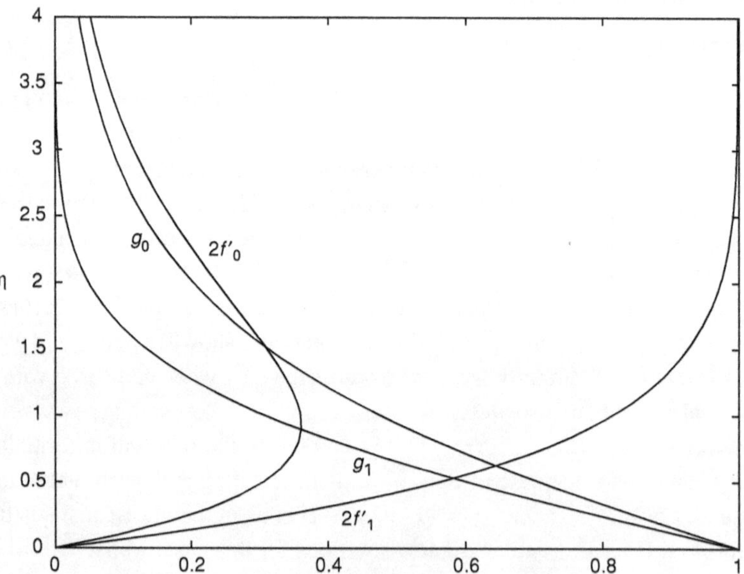

Figure 3.15 As for figure 3.14 except that now the case $i = 1$ corresponds to the case of flow towards the rotating disk, with $\lambda_2 = \frac{1}{2}$, as considered by Hannah.

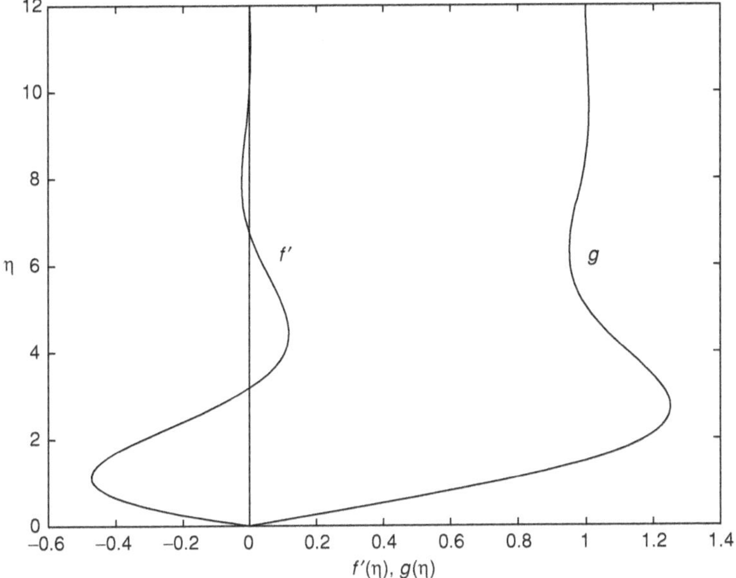

Figure 3.16 The case of solid-body rotation over a disk that is at rest; $f'(\eta)$, $g(\eta)$ are profiles of the radial and azimuthal velocity functions respectively.

axisymmetric stagnation flow, has been considered by Hall, Balakumar and Papageorgiu (1992).

Solutions with $\lambda_2 = 0$, $\lambda = s^2$, $\lambda_1 \neq 0$, $\lambda_3 = s$, have been extensively studied. Before considering these we note that the problem has attracted several rigorous approaches, and the work of McLeod (1971) deserves especial mention as the first to guarantee the existence of a solution for $s \geq 0$ and arbitrary values of λ_1, so including the classical von Kármán flow. For $s < 0$, McLeod has shown that no solution exists for $s = -1$ with $\lambda_1 = 0$ whilst Watson (1966) has established existence for $s = -1$ and λ_1 sufficiently large.

The first systematic study of equations (3.41) to (3.43) with $s \neq 0$ was carried out by Rogers and Lance (1960). With $\lambda_1 = 0$ they showed that as s increases from the classical solution with $s = 0$ the flow approaches the solid-body rotation at infinity in an oscillatory manner, with the exception $s = 1$ of course, when the whole flow is a solid-body rotation. As $s \to \infty$ the problem, suitably scaled, approaches that considered by Bödewadt where the disk is at rest in a fluid that rotates with uniform angular velocity at large distances from it. More directly, in terms of the formulation above, Bödewadt's problem is posed in equations (3.41) to (3.43) with $\lambda_1 = \lambda_2 = 0$, $\lambda = \lambda_3 = 1$ and the condition on g at $\eta = 0$ replaced by $g(0) = 0$. The solution for this case is shown in figure 3.16.

For $s < 0$ the situation proved to be more complex. The results obtained by Rogers and Lance suggested that in the absence of suction no solutions exist for $s \leq -0.2$. A subsequent investigation was carried out by Evans (1969) for $s < 0$. He was unable to obtain solutions in the range $-1.35 < s < -0.161$ in the absence of suction, but the introduction of sufficient suction did enable solutions to be calculated. A more detailed analysis has been carried out by Zandbergen and Dijkstra (1977) for $\lambda_1 = 0$ as $s \to -0.16+$. They found that the critical value of s at which the solution breaks down is $s_{cr} = -0.16053876$, where the solution $f(\eta; s)$ exhibits square-root singular behaviour such that $\partial f/\partial s \to \infty$ as $s \to s_{cr}$. They further unveiled a second solution branch which merges with the first at $s = s_{cr}$. This second branch was continued up to $s = 0.07452563$ where a third branch appeared. Subsequently Dijkstra and Zandbergen (1978) discovered an infinity of solution branches oscillating about $s = 0$. A representation of these may be found in Zandbergen (1980), or the comprehensive review article by Zandbergen and Dijkstra (1987). At $s = 0$, say, a characteristic feature of these solutions is associated with the following:

$$f(\eta) = -\alpha \sin^2\{\beta(\eta - \eta_0)\} - s/4\alpha\beta^2, \quad g = \pm 2\beta f. \qquad (3.44)$$

It may be verified that for $\alpha \gg \beta$, $\alpha\beta \gg 1$, (3.44) represents a leading-order inviscid solution of equations (3.41) and (3.42), in the nature of a 'hump'. For each new solution branch Dijkstra and Zandbergen (1978) show that a new hump is added to the solution, four times larger than the last one on the preceding branch. Dijkstra (1980) analyses the viscous layers between successive humps.

This multiplicity of solutions in the neighbourhood of $s = 0$ does not exhaust the possibilities. Bodonyi (1975) calculated solutions for large negative s, and with increasing s showed that a breakdown of the solution occurs as $s \to -1.436$. Prior to that Ockendon (1972), allowing small values of suction ($\lambda_1 \ll 1$), constructed solutions in the range $|s| < 1.436$. Her results, which included multiple solutions again characterised by thick inviscid layers separated by thin viscous layers, suggested that as $\lambda_1 \to 0$ both of the values $s = \pm 1.436$ are rather special. Indeed whole families of solutions in the neighbourhood of $s = \pm 1.436$ have been constructed by Zandbergen (1980) for $\lambda_1 = 0$, and a summary representation is presented both there and by Zandbergen and Dijkstra (1987).

The above discussion has centred solely on axisymmetric flows. However, Hewitt, Duck and Foster (1999) have discovered that a non-axisymmetric solution is available, associated with the still axisymmetric boundary conditions. With

$$v_r = r\Omega\{f'(\eta) + \phi(\eta)\cos 2\theta\}, \quad v_\theta = r\Omega\{g(\eta) - \phi(\eta)\sin 2\theta\},$$
$$v_z = -2(\nu\Omega)^{1/2}f(\eta),$$

where η is as before, and the pressure is given by equation (3.40) with $\lambda = s^2$, we have from equations (1.19) and (1.20),

$$f''' + 2ff'' - f'^2 - \phi^2 + g^2 = s^2,$$

$$g'' + 2(fg' - f'g) = 0, \quad \phi'' + 2(f\phi' - f'\phi) = 0,$$

together with boundary conditions

$$f(0) = f'(0) = \phi(0) = 0, \quad g(0) = 1, \quad f'(\infty) = \phi(\infty) = 0, \quad g(\infty) = s.$$

A numerical investigation has yielded non-axisymmetric solutions for values of s on the interval $-0.14485 \le s \le 0$. Hall, Balakumar and Papageorgiu (1992) have also investigated non-axisymmetry of the above type. For the most part they are concerned with unsteady flow, as discussed in chapter 5.

Axial symmetry of the flow is also compromised if there is a non-axisymmetric flow over the disk. Rott and Lewellen (1967) have considered, in particular, cases in which there is a uniform stream over the rotating plane, or the plane is in pure translation adjacent to a rotating fluid; Wang (1989b) considers the case of a shear flow over a rotating plane.

Whilst the exact solutions described above relate to the rotation of an infinite disk they do find application to a finite disk of radius a, say, as long as radial outflow is maintained. In that case, if $R_a = \Omega a^2/\nu \gg 1$, the effect of the edge is localised to its neighbourhood and the flow otherwise is as described by the above exact solutions. Advantage has been taken of this by, for example, Langlois (1985) and Riley (1987) in their models of the crystal/melt interface associated with the Czochralski crystal growth process.

3.5.2 The two-disk problem

For the problem of two disks, separated by a distance h, say, the representations (3.39) and (3.40) may still be used, but the boundary conditions are now applied at $\eta = (\Omega/\nu)^{1/2}h = R^{1/2}$ where $R = \Omega h^2/\nu$ is the Reynolds number. It proves convenient to write $\eta = R^{1/2}\zeta$, $f = R^{1/2}F$, $g = G$ so that equations

(3.41), (3.42) become, with a prime now denoting differentiation with respect to ζ,

$$R^{-1}F''' + 2FF'' - F'^2 + G^2 = \lambda, \tag{3.45}$$

$$R^{-1}G'' + 2(FG' - F'G) = 0, \tag{3.46}$$

together with, for impermeable disks,

$$F(0) = F'(0) = 0, \quad G(0) = 1, \quad F(1) = F'(1) = 0, \quad G(1) = s, \tag{3.47}$$

and the pressure is given by

$$\frac{p - p_0}{\rho} = \frac{1}{2}\lambda r^2 \Omega^2 - 2\nu\Omega(RF^2 + F'). \tag{3.48}$$

There are six boundary conditions, (3.47), for the fifth-order system (3.45), (3.46), but as with other confined systems that have been encountered, the pressure is only determined to within the unknown constant λ which emerges from the solution procedure.

The first serious attempts to understand the flow were made by Batchelor (1951) and Stewartson (1953). For the case $s \geq 0$ Batchelor argued that for $R \gg 1$, the fluid in the core would rotate with constant angular velocity, with boundary layers at each disk. The latter would be as in the appropriate one-disk family of solutions. For counter-rotating disks with $s = -1$, Batchelor argued that in the core, when $R \gg 1$, there would be two regions of uniform, counter-rotating flow, separated by a shear layer at $\zeta = 1/2$, with boundary layers at each disk. However, Stewartson (1953) challenged these conjectures for $s = 0, -1$ and predicted that the fluid in the core would not rotate at all for large R, with boundary layers at each disk.

These matters remained unresolved until some light was shed on them by Lance and Rogers (1962). For $s = 0, 0.5$ numerical solutions of equations (3.45) to (3.47) were obtained up to $R = O(10^2)$. As R increases the flow development in each case is the same. Boundary layers develop on each disk with inflow on the slower, outflow on the faster, which flank a core flow of uniform angular velocity as predicted by Batchelor. The solutions obtained are consistent with single-disk solutions at each disk, with the same axial velocity at the edge of the boundary layer. For $s = -0.3$ the solutions obtained by Lance and Rogers again show an emerging core flow with uniform angular velocity, but now the emerging double boundary-layer structure on the faster rotating disk precludes the possibility of a construction from the single-disk solutions. In the case where the disks have equal and opposite angular velocities, $s = -1$, solutions were obtained up to $R = O(10^3)$. The flow which is emerging has the main body of fluid at rest except for a slow, uniform inward drift to supply the slow axial motion from the plane of symmetry towards the disks, essentially as proposed by Stewartson. Subsequent rigorous analysis by McLeod and Parter (1974) established the existence of a Stewartson-type solution for $s = -1$, but throws no light on uniqueness or otherwise. Meanwhile Pearson (1965), by integrating their unsteady analogues, achieved steady-state solutions of equations (3.45) to (3.47) for $s = 0$ and $R = 1000$ confirming the existence of a Batchelor-type solution. But for the counter-rotating case $s = -1$ at $R = 1000$ a solution which is not antisymmetric about the centre-line $\zeta = 1/2$ emerges, and is therefore different from both the Batchelor–Stewartson proposals, and the calculated solutions of Lance and Rogers. The structure of the solution, which has been further investigated by Tam (1969), provides evidence for emerging inviscid structures of the type shown in equation (3.44). This evidence of non-uniqueness from the work of Pearson was followed by the landmark paper of Mellor, Chapple and Stokes (1968) for the case $s = 0$. For $R \gtrsim 200$ they not only found numerical solutions of both Batchelor and Stewartson type, but in addition solutions incorporating two or three cells of the form shown in equation (3.44). Some years later Kreiss and Parter (1983), by rigorous analytical means, established existence and non-uniqueness of solutions for all s when R is sufficiently large. Meanwhile several authors were extending the earlier solutions of Lance and Rogers to higher values of R. For example Pesch and Rentrop (1978) calculated the counter-rotating case $s = -1$ up to $R = 20000$ with the Stewartson solution clearly emerging. For $s = 0$, Wilson and Schryer (1978), with values of the Reynolds number up to $R = 10000$, show clearly a Batchelor-type solution with a now extensive region in the core where the angular velocity is uniform with value 0.3131Ω. Wilson and Schryer also allow suction at the rotating disk; the Batchelor-type solution persists with a uniform angular velocity in the core which increases with the suction rate. Keller and Szeto (1980) calculate the flow for $-1 < s < 1$ up to $R = 1000$. They conclude, in particular, that the solution is unique for $R \lesssim 55$.

The emergence of hump-like structures of the form (3.44) in the multiple solutions of Mellor *et al.* suggests, from experience with the single disk, that the complete solution space is very complex. This is highlighted by the work of Holodniok, Kubicek and Hlavacek (1981) who carried out calculations at $R = 625$ for $|s| \leq 1$. The relationship between the constant λ in equation (3.45) and s, also reproduced by Zandbergen and Dijkstra (1987), shows that no fewer than twenty solution branches have been identified.

In the above situations both disks rotate co-axially around the common axis $r = 0$, about which the flow has been assumed symmetrical. Unexpectedly, an exact solution of the Navier–Stokes equations is also available between infinite parallel planes that each rotate with angular velocity Ω about different axes. This was identified by Berker (1963), rediscovered by Abbot and Walters (1970), and further discussed by Berker (1982).

With the disk at $z = 0$ rotating about $x = 0$, $y = -l$, and that at $z = h$ about $x = 0$, $y = l$ the boundary conditions require, at $z = 0$,

$$v_r = -\Omega l \cos\theta, \quad v_\theta = \Omega(r + l\sin\theta), \quad v_z = 0, \tag{3.49}$$

and at $z = h$,

$$v_r = \Omega l \cos\theta, \quad v_\theta = \Omega(r - l\sin\theta), \quad v_z = 0. \tag{3.50}$$

The nature of the boundary conditions suggests a solution of the form

$$v_r = \Omega l\{f(\eta)\cos\theta + g(\eta)\sin\theta\},$$
$$v_\theta = \Omega r + \Omega l\{g(\eta)\cos\theta - f(\eta)\sin\theta\}, \quad v_z = 0, \tag{3.51}$$

where $\eta = z/h$, which satisfies the continuity equation identically. With the pressure given by $p - p_0 = \rho\Omega^2 r^2/2$, equations (1.19), (1.20) yield the following, linear, equations for f and g,

$$f'' + Rg = 0, \quad g'' - Rf = 0, \quad \text{where again} \quad R = \Omega h^2/\nu, \tag{3.52}$$

together with, from (3.49) and (3.50), the boundary conditions

$$f(0) = -1, \quad g(0) = 0, \quad f(1) = 1, \quad g(1) = 0. \tag{3.53}$$

The solutions of equations (3.52), subject to (3.53), with $\alpha = (R/2)^{1/2}$, are

$$\begin{aligned}
f = {} & [\{\sinh\alpha\eta\cos\alpha\eta + \sinh\alpha(\eta-1)\cos\alpha(\eta-1)\}\sinh\alpha\cos\alpha \\
& + \{\cosh\alpha\eta\sin\alpha\eta + \cosh\alpha(\eta-1)\sin\alpha(\eta-1)\}\cosh\alpha\sin\alpha]/ \\
& (\sinh^2\alpha\cos^2\alpha + \cosh^2\alpha\sin^2\alpha),
\end{aligned}$$

$$\begin{aligned}
g = {} & [\{\cosh\alpha\eta\sin\alpha\eta + \cosh\alpha(\eta-1)\sin\alpha(\eta-1)\}\sinh\alpha\cos\alpha \\
& - \{\sinh\alpha\eta\cos\alpha\eta + \sinh\alpha(\eta-1)\cos\alpha(\eta-1)\}\cosh\alpha\sin\alpha]/ \\
& (\sinh^2\alpha\cos^2\alpha + \cosh^2\alpha\sin^2\alpha).
\end{aligned}$$

The solution implies that each plane $\eta = \eta_0$, $0 < \eta_0 < 1$ rotates as if rigid, with angular velocity Ω, about an axis $x_0 = -lg(\eta_0)$, $y_0 = lf(\eta_0)$. The locus of such points is a straight line only in the limit $R = 0$. In figure 3.17 we show the projection of this locus on the (x, y)-plane for various values of the Reynolds number.

Rotating-disk flows find application in viscometry. Abbot and Walters were concerned, in particular, with a theoretical understanding of the then recently

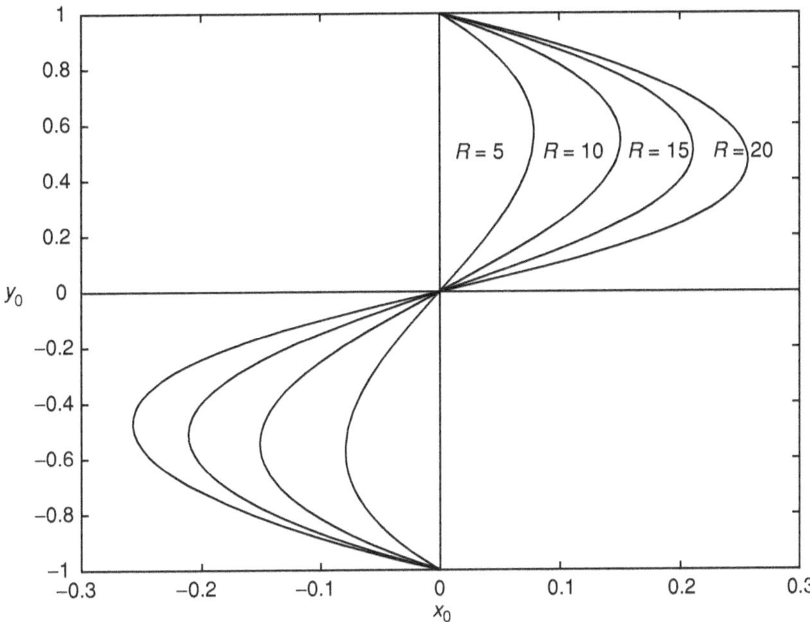

Figure 3.17 Projection onto the (x, y)-plane of the locus of points in $0 \leq z/h \leq 1$, for various values of R, about which rigid rotation takes place, between off-set rotating planes at $z/h = 0, 1$.

introduced orthogonal rheometer to determine the complex viscosity of an elastico-viscous liquid.

3.6 Ekman flow

A flow not unrelated to the above involves streaming across a single rotating plane in an otherwise unbounded fluid, which itself is in almost solid-body rotation with angular velocity $\boldsymbol{\Omega} = \Omega \mathbf{k}$. If the Navier–Stokes equations are written in a frame of reference that rotates with that same angular velocity then, if $\mathbf{v}^*$ now denotes fluid velocity relative to the rotating frame we have, for steady flow,

$$(\mathbf{v}^*.\nabla)\mathbf{v}^* + 2\boldsymbol{\Omega} \wedge \mathbf{v}^* = -\frac{1}{\rho}\nabla p^* + \nu\nabla^2\mathbf{v}^*, \quad \nabla.\mathbf{v}^* = 0, \tag{3.54}$$

where in (3.54) p^* now denotes the reduced pressure incorporating as it does the centrifugal term $\boldsymbol{\Omega} \wedge (\boldsymbol{\Omega} \wedge \mathbf{r})$. If in this rotating frame $\mathbf{v}^* = (u_\infty^*, v_\infty^*, 0)$ is constant at large distances from the plane boundary then the solution of

(3.54) will be of the form $\mathbf{v}^* = \{u^*(z), v^*(z), 0\}$ with $p^* = p^*(x, y)$ such that $\partial p^*/\partial x = 2\rho\Omega v_\infty^*$, $\partial p^*/\partial y = -2\rho\Omega u_\infty^*$ and u^*, v^* satisfying

$$\nu\frac{\partial^2 u^*}{\partial z^2} + 2\Omega(v^* - v_\infty^*) = 0, \quad \nu\frac{\partial^2 v^*}{\partial z^2} - 2\Omega(u^* - u_\infty^*) = 0. \tag{3.55}$$

In the rotating frame of reference $u^* = v^* = 0$ at $z = 0$ and the solution of (3.55) is, then,

$$u^* = u_\infty^* - \mathrm{e}^{-\eta}(u_\infty^* \cos\eta + v_\infty^* \sin\eta), \quad v^* = v_\infty^* - \mathrm{e}^{-\eta}(v_\infty^* \cos\eta - u_\infty^* \sin\eta),$$

where $\eta = (\Omega/\nu)^{1/2}z$. The velocity $\mathbf{v}^*$ so determined moves on a spiral as η increases known as the Ekman spiral following the pioneering work of Ekman (1905) in the field of rotating fluid flow dynamics.

3.7 Concentrated flows: jets and vortices

3.7.1 The round jet

Jets are characterised by the momentum flux within them stimulated, for example, by the application of a force at some given point. If M_c is a component of momentum flux across a sphere of radius r then M_c/ρ has dimensions $L^4 T^{-2}$, the only other parameter involved, the kinematic viscosity ν, has dimensions $L^2 T^{-1}$ and so, since no natural length or time scales are involved, a solution of similarity form may be anticipated. This, rather than the more obvious flow in a cone, may be considered as the analogue of the Jeffery–Hamel flow discussed in chapter 2.

Using spherical polar co-ordinates, and assuming the flow is independent of ϕ, the stream function may be written as

$$\psi = \nu r f(\eta) \quad \text{where} \quad \eta = \cos\theta, \tag{3.56}$$

so that the velocity components are

$$v_r = \frac{\nu}{r} f'(\eta) \quad \text{and} \quad v_\theta = \frac{\nu}{r}\frac{f(\eta)}{(1 - \eta^2)^{1/2}}. \tag{3.57}$$

If the pressure is written as

$$\frac{p - p_0}{\rho} = \frac{\nu^2}{r^2} P(\eta), \tag{3.58}$$

then introducing (3.57) and (3.58) into equation (1.26) yields for P, following integration,

$$P = -\frac{1}{2}\frac{f^2}{1 - \eta^2} + f' + k_1, \tag{3.59}$$

where k_1 is a constant of integration. Substituting (3.57), (3.58) and (3.59) into equation (1.25) and integrating twice gives, as equation for f,

$$\tfrac{1}{2}(1 - \eta^2)f' + \tfrac{1}{4}f^2 + \eta f + \tfrac{1}{2}k_1\eta^2 + k_2\eta + k_3 = 0, \qquad (3.60)$$

where k_2, k_3 are arbitrary constants.

Yatseyev (1950) has obtained the general solution of equation (3.60), an equation first derived by Slezkin (1934). Application to the round jet was first discussed by Landau (1944), and subsequently in more detail by Squire (1951, 1952, 1955). The first point to note is that if $\eta = 1$, that is $\theta = 0$, is taken as the jet axis then $f(1) \equiv 0$ and so if $f'(1)$, and hence the radial velocity on the axis, are to be finite it is necessary to set $k_1 = -k_2 = 2k_3 = -2k$, say, so that the last three terms of (3.60) become $-k(1 - \eta)^2$. Several special cases are worthy of attention.

(i) $k = 0$. In this case the solution is, with a an arbitrary constant,

$$f = -\frac{2(1 - \eta^2)}{a + 1 - \eta}, \quad \text{so that} \quad f' = \frac{4a\eta - 2(\eta - 1)^2}{(a + 1 - \eta)^2}, \qquad (3.61)$$

from which we have $v_r|_{\theta=0} = 4v/ar$. The solution (3.61) is appropriate to a round jet in an unbounded fluid due to a point force, that is a source of momentum, at the origin. By considering the flux of x-momentum, M_x, across a sphere of radius r Squire (1951) has shown that $a = a(M_x)$ may be calculated from

$$\frac{M_x}{2\pi\rho v^2} = \frac{32(1 + a)}{3a(2 + a)} + 8(1 + a) - 4(1 + a)^2 \ln\left(\frac{2 + a}{a}\right),$$

so that, in particular, as $a \to 0$,

$$\frac{M_x}{2\pi\rho v^2} \sim \frac{16}{3a} + 4\ln a.$$

Streamline patterns for this flow are shown in figure 3.18.

It is interesting to note that the theoretical results for the round laminar jet agree well with the experimental results for a turbulent jet, see Hinze and van der Hegge Zijnen (1949). For the turbulent jet v must be taken as an eddy viscosity, and the good agreement suggests that for jet-like flows the assumption of a constant eddy viscosity is a good one.

(ii) $k = -(1 + 4b^2)/4$. For this second case, considered by Squire (1952), the boundary condition $f = 0$ is imposed at both $\eta = 1$ and $\eta = 0$ so that the plane $\theta = \pi/2$ may be considered as a solid boundary. The appropriate solution is

$$f = -\frac{(1 + 4b^2)(1 - \eta)}{2b\cot\{b\ln(1 + \eta)\} - 1},$$

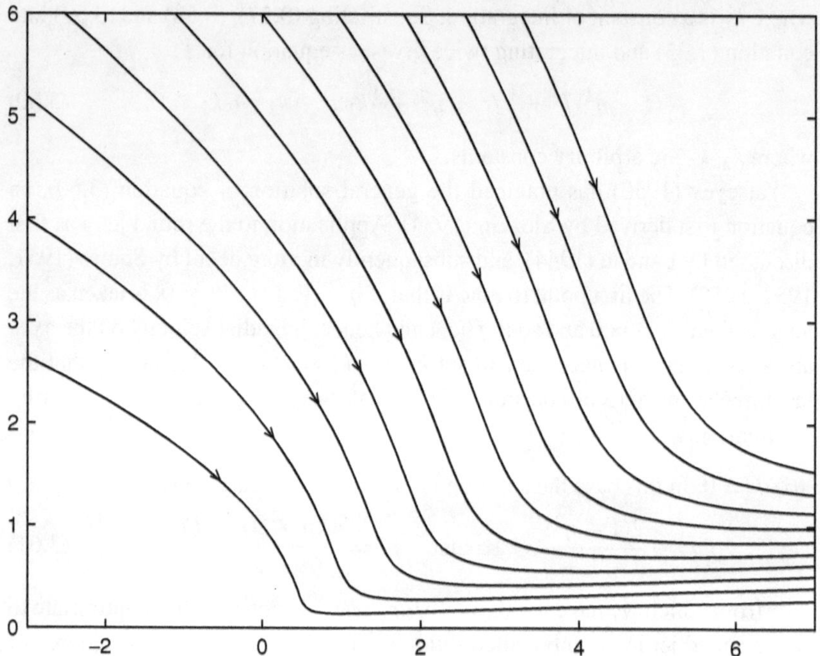

Figure 3.18 Streamlines for an unbounded round jet due to a point source of momentum, from equation (3.61) with $a = 0.01$.

so that

$$f' = \frac{1 + 4b^2}{2b \cot\{b \ln(1 + \eta)\} - 1} \left[1 - 2b^2 \left(\frac{1 - \eta}{1 + \eta} \right) \frac{\mathrm{cosec}^2\{b \ln(1 + \eta)\}}{2b \cot\{b \ln(1 + \eta)\} - 1} \right],$$

with

$$f'(1) = \frac{1 + 4b^2}{2b \cot(b \ln 2) - 1}, \quad f'(0) = -\frac{1}{2}(1 + 4b^2).$$

Thus, although $v_\theta = 0$ on $\theta = \pi/2$, $v_r \neq 0$ and the solution suffers from the drawback that it does not satisfy the no-slip condition at the plane boundary from which the jet originates at $r = 0$. Narrow high-speed jets correspond to values of b for which $2b \cot(b \ln 2) \approx 1$; an example is shown in figure 3.19.

(iii) $k_1 = k_2 = 0$, $k_3 = 1 - b^2$. This case, unlike the two previous examples, does not suppress the singular behaviour of the solution. From equation (3.60) we now have

$$f = 2 \left\{ \frac{(1 + \eta)^b (b - \eta) - c(1 - \eta)^b (b + \eta)}{(1 + \eta)^b + c(1 - \eta)^b} \right\},$$

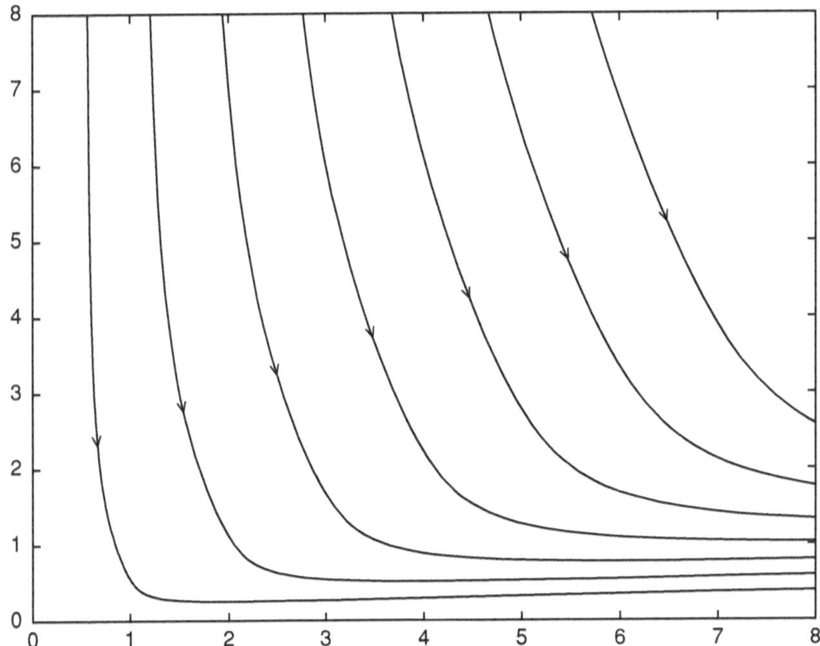

Figure 3.19 As in figure 3.18 except that the flow is bounded by the plane wall $\theta = \frac{1}{2}\pi$. The case shown has $b = 1.88$.

where c is a constant. Berker (1963), making a small correction to Squire's (1955) choice of c, gives

$$c = \left(\frac{1 + \eta_0}{1 - \eta_0}\right)^b \left(\frac{b - \eta_0}{b + \eta_0}\right),$$

a choice which ensures that $v_\theta = 0$ at $\theta = \theta_0 = \cos^{-1} \eta_0$. It is, therefore, tempting to assume that this solution might represent the flow issuing from the annulus formed by two cones with the same apex and near-identical apex angles. But v_θ is unbounded at the axis $\theta = 0, \pi$, which is, in fact, a line source of strength $m = 4\pi \nu(b - 1)$. All the fluid entrained from within this conical, annular jet originates from the line source, which therefore lends a somewhat more artificial air to the solution.

Agrawal (1957) seeks to generalise (3.56) by writing

$$\psi = K r^n f(\eta)$$

where K is an arbitrary constant. In addition to the solution for $n = 1$, Agrawal recovers the solution for a uniform stream for which $n = 2$, and finds an additional solution for $n = 4$ with

$$f(\eta) = c_1(1 - \eta^2) + c_2(1 - \eta^2)(1 - 5\eta^2)$$
$$+ c_3 \left\{ (1 - \eta^2)(3 - 15\eta^2) \ln \left(\frac{1 + \eta}{1 - \eta} \right) + 30\eta^3 - 26\eta \right\},$$

where c_1, c_2 and c_3 are arbitrary constants. If singular behaviour at the axis is to be avoided then $c_3 = 0$. A particular example (already encountered in section 3.3) with $c_1 = -c_2 = 1/5$ yields

$$\psi = Kr^4\eta^2(1 - \eta^2). \tag{3.62}$$

The flow represented by the stream function (3.62) is a rotational flow, vorticity

$$\boldsymbol{\omega} = 2Kr(1 - \eta^2)^{1/2}\hat{\boldsymbol{\phi}},$$

with streamlines that are rectangular hyperbolae, against the plane boundary $\theta = \pi/2$, that is the (x, y)-plane, at which $\mathbf{v} = \mathbf{0}$.

Morgan (1956) has considered solutions of equation (3.60) that represent the flow bounded by either a single cone, or a pair of cones with a common apex. Although the no-slip condition is satisfied at the boundaries, non-trivial solutions are only available when there is transpiration of fluid across them.

3.7.2 The Burgers vortex

The inviscid flow, in a cylindrical polar co-ordinate system,

$$v_r = -kr, \quad v_z = 2kz, \tag{3.63}$$

formed the basis for the earlier study in this chapter of the stagnation flow on a circular cylinder. Suppose the rigid circular cylinder is replaced by a cylindrical vortex core, which induces a circulation Γ at large distances. This suggests a form for the azimuthal velocity as

$$v_\theta = \frac{\Gamma}{2\pi r} g(r), \quad \text{with} \quad g(\infty) = 1. \tag{3.64}$$

Substitution of (3.63), (3.64) into equation (1.20) gives, as the equation for g,

$$r\frac{d^2 g}{dr^2} + \left(\frac{r^2 k}{\nu} - 1 \right) \frac{dg}{dr} = 0,$$

with solution

$$g = 1 - e^{-\eta}, \quad \text{where} \quad \eta = \frac{kr^2}{2\nu}, \tag{3.65}$$

and the choice $g(0) = 0$ ensures $v_\theta = 0$ at $r = 0$. The vorticity within the core is given by $\boldsymbol{\omega} = (\Gamma/2\pi\nu)e^{-\eta}\hat{\mathbf{z}}$. This solution discovered by Burgers (see Burgers (1948)), epitomises the dynamics of vorticity discussed in chapter 1, as indeed does the two-dimensional analogue of Robinson and Saffman (1984). The axial velocity v_z stretches the vortex tubes within the core of vorticity and so intensifies it. The vorticity readily diffuses under the action of viscosity, but is restrained by the inward radial convection of vorticity due to the radial velocity v_r.

Sullivan (1959) has extended Burgers' result. If the velocity components are written as

$$v_r = -kr + \frac{1}{r} f(\eta), \quad v_z = 2kz \left\{ 1 - \frac{f'(\eta)}{2\nu} \right\}, \quad v_\theta = \frac{\Gamma}{2\pi r} \frac{g(\eta)}{g(\infty)}, \quad (3.66)$$

then Sullivan shows that

$$f(\eta) = 6\nu(1 - e^{-\eta}), \quad g(\eta) = \frac{\nu}{k} \int_0^\eta \exp\left[-t + 3 \int_0^t \{(1 - e^{-s})/s\}\, ds \right] dt. \quad (3.67)$$

The structure of the vortical core differs significantly from that of Burgers. For a Burgers vortex $v_r < 0$, $v_\theta > 0$, $v_z/z > 0$ for all r, but from (3.66), (3.67) it can be shown that whilst $v_\theta > 0$ for all r, albeit with a different distribution from (3.64), (3.65), $v_r > 0$ for $\eta \lesssim 2.821$ and $v_z/z < 0$ for $\eta \lesssim 1.099$. As a consequence, whilst the flow assumes the character of a Burgers vortex for $\eta > 2.821$, there is an inner 'cell' within which v_z/z changes sign resulting in a region of counterflow near the axis. The flow therefore assumes, as Sullivan remarks, a two-cell structure.

3.7.3 The influence of boundaries

The Burgers vortex is atypical of real vortices, in the sense that no boundaries are present. Whilst in the solution (3.64), (3.65) k and Γ may be specified independently, the presence of a rigid boundary couples the swirling and, induced, secondary motion.

Long (1958, 1961b), working with cylindrical polar co-ordinates, showed that the Navier–Stokes equations reduce to coupled ordinary differential equations for a conical flow with a line vortex on the half-line $r = 0$, $z \geq 0$. In his solution the circulation Γ about the axis tended to a constant, K, as $r \to \infty$, but the solution did not satisfy $\mathbf{v} = \mathbf{0}$ on the boundary $z = 0$. Whilst the solution is well behaved in the upper half-plane it is singular at $r = z = 0$. As Long observed, his solution can only be expected to be valid at large distances from the bounding surface $z = 0$.

Goldshtik (1960) and Serrin (1972) overcame the difficulty of the bounding surface by developing a solution for which $\mathbf{v} \equiv \mathbf{0}$ on $z = 0$, but at the expense of introducing singular behaviour along the axis. Indeed it appears, as with the round jet discussed above, that for these conical flows it is not possible to have boundaries on which $\mathbf{v} = \mathbf{0}$ without introducing unwanted singular behaviour elsewhere.

As with the round jet it is convenient to use spherical polar co-ordinates, and to supplement (3.56), (3.57) by writing the azimuthal, or swirling, component of velocity and pressure as

$$v_\phi = \frac{C}{r}\frac{\Omega(\eta)}{(1-\eta^2)^{1/2}}, \qquad \frac{p - p_0}{\rho} = \frac{P(\eta)}{r^2(1-\eta^2)} \tag{3.68}$$

where C is related to the circulation. From the radial momentum equation (1.25) the pressure function $P(\eta)$ is given by

$$-2P = f^2 + k^2\Omega^2 + \{ff'' + f'^2 + (1-\eta^2)f''' - 2\eta f''\}(1-\eta^2) \tag{3.69}$$

where $k = C/\nu$ is essentially a Reynolds number. Equations (1.26) and (1.27) then give

$$(1-\eta^2)f^{iv} - 4\eta f''' + ff''' + 3f'f'' = -2k^2\Omega\Omega'/(1-\eta^2), \tag{3.70}$$

and

$$(1-\eta^2)\Omega'' + f\Omega' = 0. \tag{3.71}$$

If the no-slip condition is to be satisfied at the bounding plane $z = 0$, that is $\eta = 0$, then

$$f(0) = f'(0) = \Omega(0) = 0; \tag{3.72}$$

whilst if the vortex, aligned with the axis, is to be neither a line source nor a sink, then

$$f \to 0, \quad \Omega \to 1 \quad \text{as} \quad \eta \to 1, \tag{3.73}$$

corresponding to a line vortex of circulation $\Gamma = 2\pi C$.

Equation (3.70) may be integrated three times to give, with some manipulation and using conditions (3.72)

$$2(1-\eta^2)f' + 4\eta f + f^2 = -2k^2 \int_0^\eta \frac{(\eta - t)(1 - \eta t)}{(1-t^2)^2} \Omega^2 \, dt + k^2(\tilde{P}\eta^2 + \tilde{Q}\eta), \tag{3.74}$$

where the notation of Serrin (1972) has, in part, been retained. Equation (3.74) may be compared with its non-swirling counterpart (3.60). The integral in equation (3.74) converges as $\eta \to 1$. Using (3.73), it may be then inferred that

$(1 - \eta)f' \to 0$ as $\eta \to 1$ so that

$$\tilde{P} + \tilde{Q} = 2 \int_0^1 \frac{\Omega^2}{(1+t)^2} dt,$$

and (3.74) may finally be written as

$$2(1 - \eta^2)f' + 4\eta f + f^2 = 2k^2(1 - \eta)^2 \int_0^\eta \frac{t\Omega^2}{(1 - t^2)^2} dt$$

$$+ 2k^2\eta \int_\eta^1 \frac{\Omega^2}{(1+t)^2} dt + k^2 \tilde{P}(\eta^2 - \eta). \tag{3.75}$$

The integro-differential system of equations (3.71) and (3.75) has been anal-ysed by Serrin (1972) who also presents numerical solutions for specific pairs $(k, \tilde{P})$ of the two parameters. But there is a difficulty. As $\eta \to 1$ Serrin shows that

$$f \sim \tfrac{1}{8}(\tilde{P} - 1)k^2(1 - \eta^2)\ln(1 - \eta),$$

and this implies

$$v_r \sim -\frac{\nu k^2}{4r}(\tilde{P} - 1)\ln(1 - \eta) \quad \text{as} \quad \eta \to 1, \tag{3.76}$$

so that not only is the azimuthal velocity component singular, as a line vortex, but so also is the axial component of velocity as $\eta \to 1$. That is, except in the case $\tilde{P} = 1$ which was the case considered by Goldshtik (1960). It has been shown by Goldshtik and Shtern (1990), (see also Goldshtik (1990)), that the singular axial flow (3.76) may be attributed to a line force on $\eta = 1$ given by

$$F_r|_{\eta=1} = \frac{\pi \rho v^2}{r}(\tilde{P} - 1). \tag{3.77}$$

As Serrin notes there is also an unbalanced radial pressure force, acting towards the axis, close to the bounding surface where the swirling component of velocity reduces to zero. He analyses properties of solutions of (3.71), (3.75) including a delineation of $(k, \tilde{P})$-parameter space where no solutions exist. For the case $\tilde{P} = 1$, considered by Goldshtik (1960) solutions only exist for $k \lesssim 5.53$. Goldshtik and Shtern (1990) analyse the behaviour of solutions as they lose existence, termed 'flow collapse', and also demonstrate non-uniquness of the solutions considered by Serrin in some parameter sub-region.

The solutions calculated by Serrin are of three distinct types. If $\tilde{P}$ is suffi-ciently small there is an axial downflow and radial outflow over the bounding plane, whilst if $\tilde{P}$ is sufficiently large, radial inflow is accompanied by axial upflow. For a range of intermediate values of $\tilde{P}$ there is both axial downflow and radial inflow compensated by outflow along a cone $\eta = \eta_0$, say, where $0 < \eta_0 < 1$. These are shown in figure 3.20.

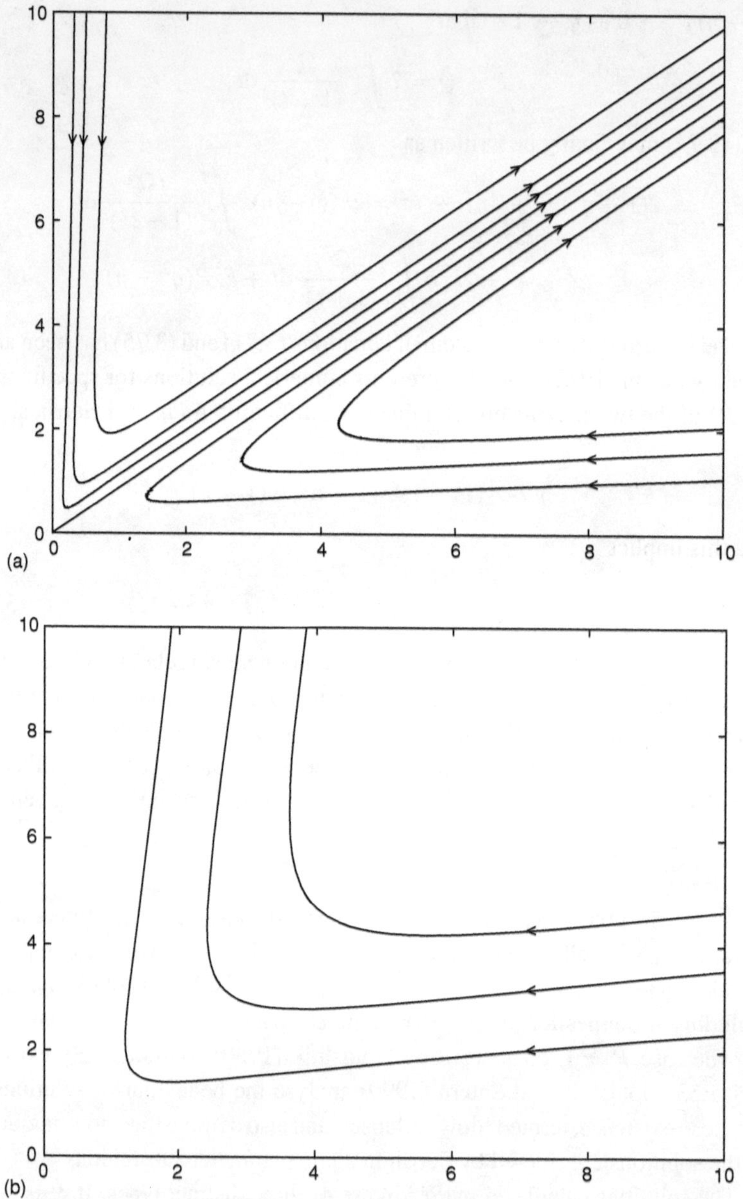

Figure 3.20 Streamline patterns for the bounded vortex configuration of Serrin. (a) $k = 5$, $\tilde{P} = 0.4$. In this case the radial inflow along the boundary and axial downdraught compete, resulting in a funnel-like flow. (b) $k = 3.0$, $\tilde{P} = 1.2$. A case in which the downdraught is overwhelmed by the radial inflow. (c) $k = 3.5$, $\tilde{P} = 0.2$. The obverse of (ii).

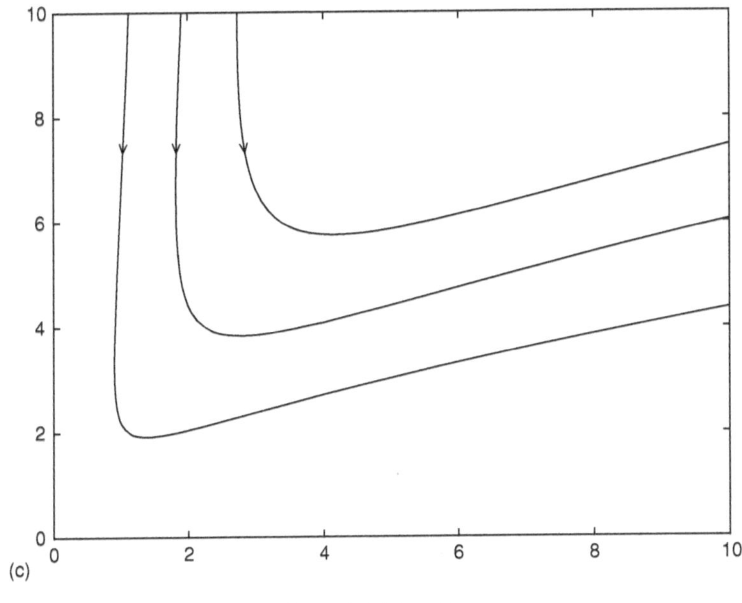

(c)

Figure 3.20 (*cont.*)

In figure 3.21 we show radial velocity profiles correponding to the three types of solution shown in figure 3.20.

Whilst refraining from a claim that his solutions represent physical phenomena, there are clearly features reflecting the behaviour of tornadoes and waterspouts; Serrin draws attention to Morton (1966) for an account both of the relation of the concentrated vortex core to the main motion and of the driving mechanisms likely to be involved in typical tornado and waterspout phenomena.

Yih, Wu, Garg and Leibovich (1982) also consider a conical vortex flow of the type represented by equations (3.56) to (3.58) and (3.68), governed by the differential equations (3.70) and (3.71). The flow is bounded by the cone $\theta = \theta_0$, with the special case $\theta_0 = \pi/2$ corresponding to that of Goldshtik (1960) and Serrin (1972). However, unlike Goldshtik and Serrin, they relax the no-slip condition at the solid boundary but insist that velocities be finite along the axis. In that case the boundary conditions require

$$f(\eta_0) = 0, \quad \Omega(\eta_0) = 1, \quad f(1) = \Omega(1) = 0, \quad \lim_{\eta \to 1}(1 - \eta^2)f''(\eta) = 0,$$

where, with C arbitrary in (3.68), the condition on Ω at the boundary establishes the level of swirl. The solution is not entirely free from singularities; there is a singularity at the cone vertex which Yih *et al.* interpret as a source of axial

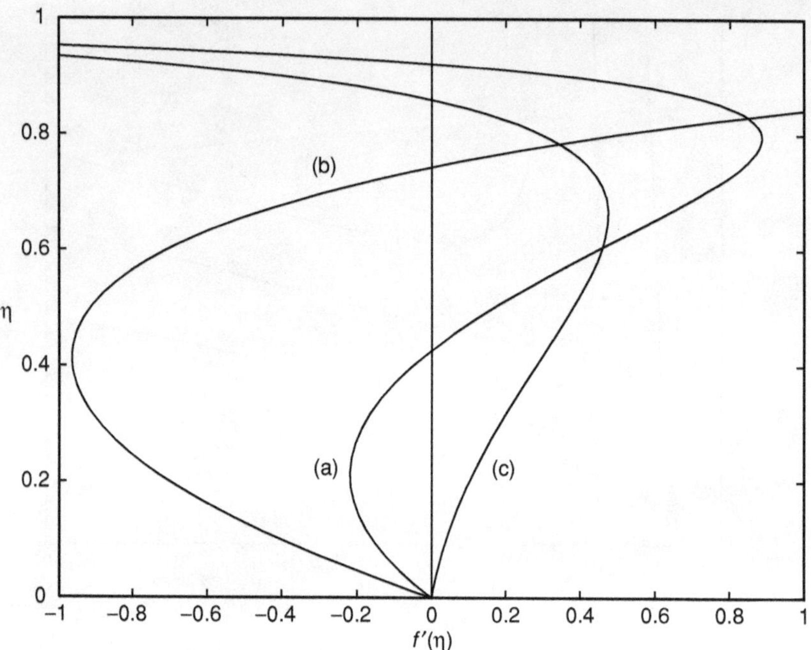

Figure 3.21 Radial velocity profiles $rv_r/v = f'(\eta)$ corresponding to the stream-line patterns of figures 3.20 (a), (b), (c).

momentum. Numerical solutions are presented for cases $\theta_0 = \pi/4, \pi/2, 3\pi/4$. For the case of a plane boundary, $\theta = \pi/2$, solutions similar to each of the three types shown in figure 3.20 are shown. In a subsequent paper Sozou (1992) extends the work of Yih *et al.* by introducing a source of vorticity at the vertex of the cone. Formal existence of the class of solutions obtained by Yih *et al.* has been established by Stein (2000). Stein (2001) has also addressed the problem of uniqueness, showing that the solutions are unique for $\theta_0 \leq \pi/2$, but not for $\theta_0 > \pi/2$ where in at least one case non-uniquness has been demonstrated.

In a series of papers Pillow and Paull (1985), Paull and Pillow (1985a, 1985b) carry out a comprehensive investigation of conically similar viscous flows. In particular they seek types of axial causes of such flows. For flows of the type studied by Serrin (1972) they identify the force (3.77), whilst for those of Yih *et al.* (1982) semi-infinite line sources of angular momentum generate such swirling flows.

4

Unsteady flows bounded by plane boundaries

The simplest unsteady flows involving plane boundaries, in which a single infinite boundary moves in its own plane, are associated with the names of Stokes (1851) and Rayleigh (1911). Consider a situation in which $\mathbf{v} = \{u(y, t), 0, 0\}$. Equation (1.12) is satisfied identically. Equations (1.9) and (1.10) then yield, respectively,

$$\frac{\partial u}{\partial t} = -\frac{1}{\rho}\frac{\partial p}{\partial x} + \nu\frac{\partial^2 u}{\partial y^2}, \tag{4.1}$$

$$\frac{\partial p}{\partial y} = 0. \tag{4.2}$$

Under the assumed conditions, the most general solution of equation (4.2) is $p = -\rho\{f(t) + xg(t)\}$, where f and g are arbitrary functions of t, so that (4.1) becomes

$$\frac{\partial u}{\partial t} = \nu\frac{\partial^2 u}{\partial y^2} + g(t); \tag{4.3}$$

differentiating this equation with respect to y, and introducing the vorticity $\boldsymbol{\omega} = (0, 0, \zeta)$ yields, as equation for ζ,

$$\frac{\partial \zeta}{\partial t} = \nu\frac{\partial^2 \zeta}{\partial y^2}. \tag{4.4}$$

Each of equations (4.3), (4.4) are analogous to the one-dimensional heat conduction equation. In (4.3) u plays the role of temperature with $g(t)$ representing an unsteady source of heat. The diffusion equation (4.4) has the vorticity ζ in the role of temperature with the equation representing diffusion of heat, though in a fluid-dynamical context diffusion of vorticity is a well established and valuable concept. We next consider solutions of equation (4.3), and its generalisation, in several physical contexts.

4.1 The oscillating plate

Consider first the classical problem of the flow induced when the infinite plane boundary $y = 0$ performs unidirectional oscillations, frequency ω, in its own plane such that $u(0, t) = U_0 \cos \omega t$ where U_0 is a constant. With $g(t) \equiv 0$ we seek a solution in which $u = U_0 e^{i\omega t} F(y)$ where, from (4.3), F satisfies

$$\nu F'' - i\omega F = 0, \tag{4.5}$$

with

$$F(0) = 1, \quad F \to 0 \quad \text{as} \quad y \to \infty. \tag{4.6}$$

The solution of (4.5) subject to (4.6) is $F(y) = \exp\{-(1 + i)(\omega/2\nu)^{1/2}y\}$ so that, taking the real part, we have

$$u = U_0\, e^{-(\omega/2\nu)^{1/2}y} \cos\{\omega t - (\omega/2\nu)^{1/2}y\}. \tag{4.7}$$

The shear stress exerted on the plate is given by

$$\mu \frac{\partial u}{\partial y}\bigg|_{y=0} = U_0(\rho\mu\omega)^{1/2} \cos(\omega t + 5\pi/4),$$

which is seen to differ in phase from the velocity of the boundary.

The solution (4.7) shows that vorticity, of alternating sign, created at the boundary propagates from it in a wave-like manner with speed $(2\omega\nu)^{1/2}$. This viscous wave decays in amplitude and the whole motion is confined to a layer of thickness $O\{(\nu/\omega)^{1/2}\}$, often referred to as the 'Stokes layer'. If the fluid is confined by an upper, stationary, boundary at $y = h$, say, then the condition as $y \to \infty$ is replaced by $F(h) = 0$ with corresponding solution

$$
\begin{aligned}
u = {} & \frac{U_0}{2(\cosh 2\lambda h - \cos 2\lambda h)}\{e^{-\lambda(y-2h)} \cos(\omega t - \lambda y) \\
& + e^{\lambda(y-2h)} \cos(\omega t + \lambda y) - e^{-\lambda y} \cos(\omega t - \lambda y + 2\lambda h) \\
& - e^{\lambda y} \cos(\omega t + \lambda y - 2\lambda h)\}
\end{aligned}
$$

where $\lambda = (\omega/2\nu)^{1/2}$.

The effect of suction on the Stokes-layer solution has been considered by both Stuart (1955) and Debler and Montgomery (1971). If we assume that $\mathbf{v} = \{u(y, t), v(y), 0\}$ then equation (1.12) shows that $\partial v/\partial y = 0$ and we infer that $v = \text{constant} = -V$, say, where V is the suction velocity. The analogue of equation (4.3), again with $g = 0$, is then seen to be, from (1.9),

$$\frac{\partial u}{\partial t} - V\frac{\partial u}{\partial y} = \nu\frac{\partial^2 u}{\partial y^2} \tag{4.8}$$

with, as before, $u(0, t) = U_0 \cos \omega t$ and $u \to 0$ as $y \to \infty$. If we seek a solution analogous to (4.7) in the form

$$u = U_0\, e^{-ay} \cos(\omega t + by), \qquad (4.9)$$

then direct substitution in equation (4.8) yields for the constants a and b

$$a = \left(\frac{\omega}{2\nu}\right)^{1/2} \left[\left\{ 1 + \left(\frac{V^2}{4\omega\nu}\right)^2 \right\}^{1/2} - \frac{V^2}{4\omega\nu} \right]^{-1/2} + \frac{V}{2\nu}, \qquad (4.10)$$

$$b = -\left(\frac{\omega}{2\nu}\right)^{1/2} \left[\left\{ 1 + \left(\frac{V^2}{4\omega\nu}\right)^2 \right\}^{1/2} - \frac{V^2}{4\omega\nu} \right]^{1/2}. \qquad (4.11)$$

Debler and Montgomery show that if a steady transverse velocity, W, is introduced such that now $\mathbf{v} = \{u(y, t), -V, w(y)\}$, then u is unchanged, as in (4.9), and the solution of equation (1.11) is the asymptotic suction profile analysed in chapter 2, namely

$$w = W(1 - e^{-Vy/\nu}).$$

The investigation by Debler and Montgomery was motivated by a particular papermaking device that oscillates laterally as a dilute mixture of water and wood pulp is carried along it. In their study they also utilised the ideas based on the Stokes solution for the case in which a liquid film is adjacent to the oscillating boundary, with a gaseous medium above it. The solution is completed by ensuring continuity of velocity and stress at the two-phase interface. Stuart was concerned with the effects of streamwise fluctuations on a steady flow speed U_1, say, that corresponds to the asymptotic suction profile. In that case the porous boundary is at rest and streamwise fluctuations $U_0 \cos \omega t$ are introduced. The solution for the velocity component $u(y, t)$ is readily seen to be

$$u = U_1(1 - e^{-Vy/\nu}) + U_0\{\cos \omega t - e^{-ay} \cos(\omega t + by)\}, \qquad (4.12)$$

where a and b are given by (4.10) and (4.11). Kelly (1965) has also included the effect of a time-dependent suction velocity.

4.2 Impulsive flows

The prototype impulsive flow arises when the infinite plane is set in motion with constant velocity in its own plane. This flow was analysed by Stokes (1851) and Rayleigh (1911). Watson (1955), seeking a generalisation of this flow, considered situations in which the boundary velocity $U(t)$ leads to solutions of equation (4.3), in the absence of the source term $g(t)$, which are self-similar. He found that there are two possibilities namely

$$U(t) = At^{\alpha}, \quad U(t) = A\,e^{\omega t}. \tag{4.13}$$

For the first of these we have

$$u = At^{\alpha} f(\eta) \quad \text{where} \quad \eta = y/2(\nu t)^{1/2}, \tag{4.14}$$

and from (4.3) we have, as the equation for f,

$$f'' + 2\eta f' - 4\alpha f = 0, \tag{4.15}$$

together with

$$f(0) = 1, \quad f \to 0 \quad \text{as} \quad \eta \to \infty. \tag{4.16}$$

The solution of equation (4.15), subject to (4.16), may be written as

$$f = 2^{2\alpha}\Gamma(\alpha + 1)g_{\alpha}(\eta) \quad \text{where}$$
$$g_{\alpha} = 2^{(1/2)-\alpha}\pi^{-1/2}\exp(-\eta^2/2)D_{-2\alpha-1}(\sqrt{2}\eta), \tag{4.17}$$

where $\Gamma(x)$ is the gamma function and $D_n(x)$ the parabolic cylinder function. The function $g_{\alpha}(\eta)$ has the properties that

$$g_{\alpha}'(\eta) = -g_{\alpha-1/2}(\eta), \quad g_0(\eta) = 1 - \text{erf}(\eta), \quad g_{\alpha}(0) = \frac{2^{-2\alpha}}{\Gamma(\alpha + 1)}. \tag{4.18}$$

In figure 4.1 we show the velocity function $f(\eta)$ for various values of α. The case $\alpha = 0$ is the classical case, first considered by Stokes, corresponding to the situation in which the plane moves with constant velocity from the initial instant.

The shear stress exerted on the plane is given by

$$\mu\frac{\partial u}{\partial y}\bigg|_{y=0} = \frac{1}{2}A(\rho\mu)^{1/2}t^{\alpha-1/2}f'(0) = -A(\rho\mu)^{1/2}t^{\alpha-1/2}\frac{\Gamma(\alpha + 1)}{\Gamma(\alpha + 1/2)},$$

and in figure 4.2 we plot $f'(0) = -2\Gamma(\alpha + 1)/\Gamma(\alpha + \frac{1}{2})$ as a function of α.

The effect of suction on this class of flows, with the wall suction $V(t) \propto t^{-1/2}$, has been considered by Hasimoto (1957).

The second possibility shown in equation (4.13) leads to a distribution of velocity $u(y, t) = A\,e^{\omega t}f(y)$ from which it readily emerges that $f = \exp\{-(\omega/\nu)^{1/2}y\}$. This solution was first discussed by Görtler (1944), and may be interpreted as a flow in which the plane is set into motion from a state of rest at $t = -\infty$.

From the first class of these solutions, we see from (4.14) that the vorticity created at the plane $y = 0$, a vortex sheet at the initial instant, diffuses without

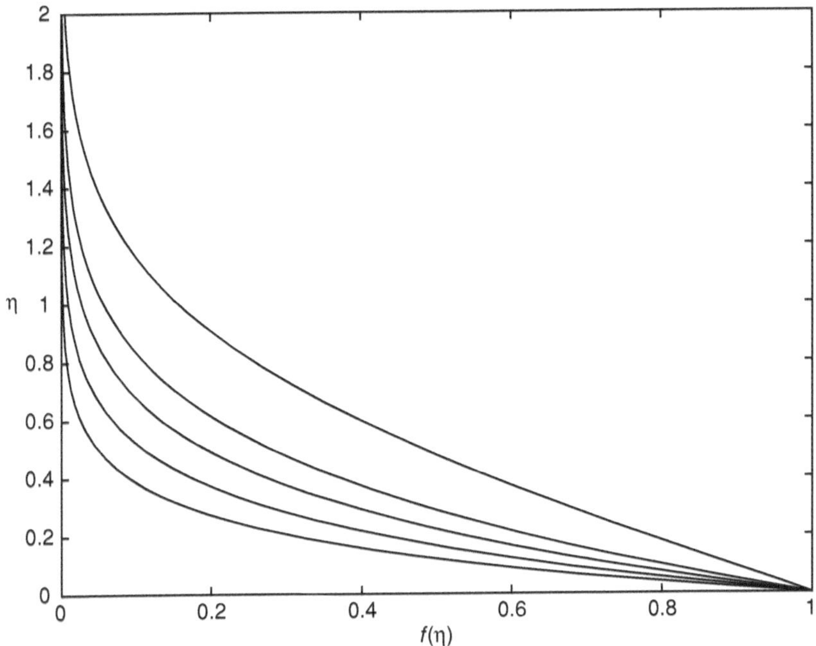

Figure 4.1 The velocity function $f(\eta)$, when the boundary moves with speed At^{α}, for values of $\alpha = 0$ (uppermost), 1, 2, 4, 8.

bound into the main body of the fluid to a distance $O\{(\nu t)^{1/2}\}$ at time t. By contrast, for the second type of solution although the vorticity intensifies at an exponential rate with respect to time it is constrained to a region of thickness $O\{(\nu/\omega)^{1/2}\}$ as in the Stokes layer.

Other planar impulsive flows are outlined by Berker (1963) including the following.

4.2.1 Applied body force

If a body force $\mathbf{F} = k\mathbf{i}$ is applied to the fluid which is at rest initially, with the boundary $y = 0$ at rest, then $u(y, t)$ satisfies equation (4.3) with $g(t)$ replaced by k, and $u(y, 0) = 0$, $0 < y < \infty$, $u(0, t) = 0$, $0 < t < \infty$. Setting $u = kt + \bar{u}$ then shows that $\bar{u}$ satisfies

$$\frac{\partial \bar{u}}{\partial t} = \nu \frac{\partial^2 \bar{u}}{\partial y^2}, \tag{4.19}$$

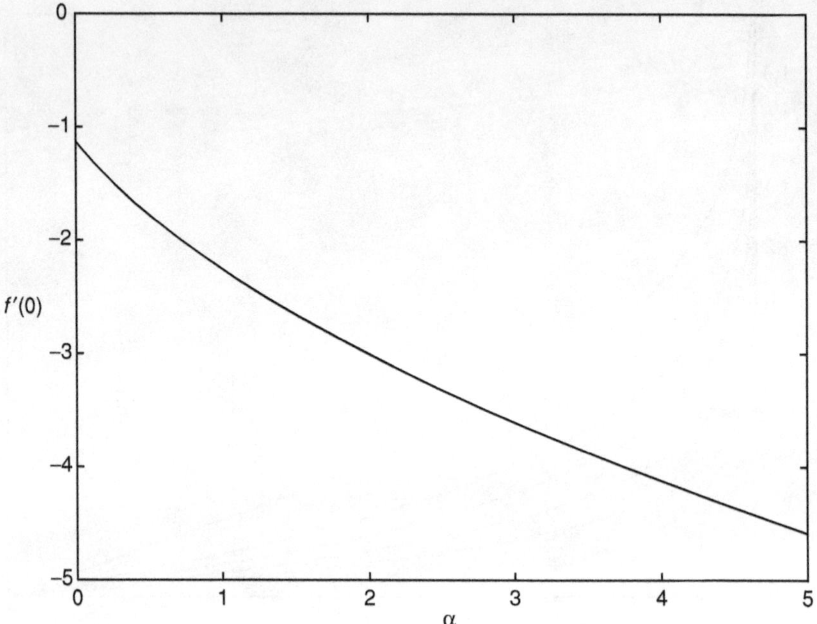

Figure 4.2 The shear-stress function $f'(0)$ for the class of flows illustrated in figure 4.1.

together with

$$\bar{u}(y, 0) = 0, \quad 0 < y < \infty, \quad \bar{u}(0, t) = -kt, \quad 0 < t < \infty;$$

$$\bar{u} \to 0 \quad \text{as} \quad y \to \infty. \qquad (4.20)$$

The problem posed by equations (4.19), (4.20) is a special case of those investigated by Watson, considered above, with $\alpha = 1$. The solution is given by (4.14), (4.17) with $A = -k$ so that, finally,

$$u = kt\{1 - 2^{-3/2}\pi^{-1/2}\exp(-\eta^2/2)D_{-3}(\sqrt{2}\eta)\}.$$

4.2.2 Applied shear stress

An alternative to inducing a motion in the bulk of the fluid by setting the plane boundary $y = 0$ into motion is to apply a shear stress at the boundary, so that $\mu\, \partial u/\partial y|_{y=0} = -T$ with, for example, T constant. If we write

$$\bar{u} = \mu\frac{\partial u}{\partial y} \qquad (4.21)$$

then $\bar{u}$ satisfies equation (4.19) and the conditions (4.20) except that the condition at $y = 0$ is replaced by $\bar{u}(0, t) = -T, 0 < t < \infty$. The problem for $\bar{u}$ is, then, the classical problem of an impulsively moved plane with constant speed. The solution we infer from (4.14) and (4.18), with $\alpha = 0$, as

$$\bar{u} = T \left\{ \frac{2}{\sqrt{\pi}} \int_0^{\eta} e^{-s^2} \, ds - 1 \right\} \quad \text{where} \quad \eta = y/2(vt)^{1/2}. \tag{4.22}$$

From this expression we deduce that

$$u = \frac{T}{\mu} \left[2 \left(\frac{vt}{\pi} \right)^{1/2} e^{-y^2/4vt} + y \left\{ \text{erf} \left(\frac{y}{2(vt)^{1/2}} \right) - 1 \right\} \right], \tag{4.23}$$

with $u(0, t) = 2T(t/\rho\mu\pi)^{1/2}$. Berker (1963) explores the possibility that the solution (4.23) may find application in the study of wind-driven marine currents.

4.2.3 Diffusion of a vortex sheet

As a preliminary we note that $u = t^{-1/2} \exp(-y^2/4vt)$ is a solution of equation (4.3) in the absence of the source term and that, therefore, so also is

$$u = \frac{1}{2(\pi vt)^{1/2}} \int_{-\infty}^{\infty} e^{-(s-y)^2/4vt} f(s) \, ds, \tag{4.24}$$

a solution that has the property $u \to f(y)$ as $t \to 0$, for $-\infty < y < \infty$. Consider then the case when the initial velocity distribution is given as

$$u = \begin{cases} u_0, & \text{for} \quad y > 0, \\ -u_0, & \text{for} \quad y < 0, \end{cases}$$

where u_0 is a constant. This distribution implies the presence of a vortex sheet at $y = 0$ at the initial instant. From equation (4.24) we then have

$$u = \frac{u_0}{2(\pi vt)^{1/2}} \left(\int_0^{\infty} e^{-(s-y)^2/4vt} \, ds - \int_0^{\infty} e^{-(s+y)^2/4vt} \, ds \right). \tag{4.25}$$

From the solution (4.25) we find the normal component of vorticity as

$$\zeta = -\frac{u_0}{(\pi vt)^{1/2}} e^{-y^2/4vt}. \tag{4.26}$$

Distributions of this vorticity, at various times, as the vortex sheet diffuses are shown in figure 4.3. The total vorticity remains constant, of course, and equal to $-2u_0$ per unit length.

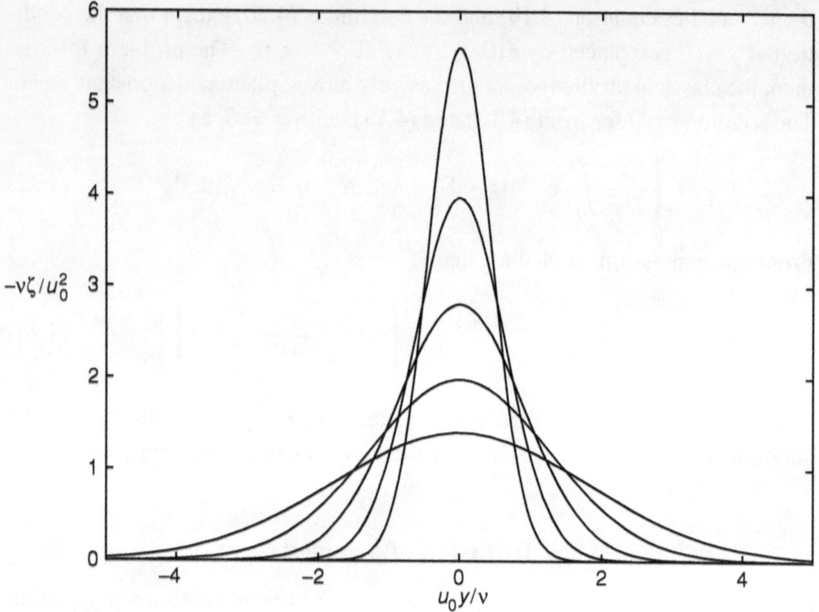

Figure 4.3 The vorticity in a diffusing vortex sheet for values of $\bar{t} = u_0^2 t/\nu = 0.1$ (uppermost), 0.2, 0.4, 0.8, 1.6.

4.3 More general flows

Watson (1958) uses the asymptotic suction profile as a steady base flow on which unsteady motions are superposed, whilst Hasimoto (1957) has considered situations, with uniform suction, for which the free stream $U(t) \equiv 0$ for $t < 0$. With $v \equiv -V$ and a free stream $U(t) = U_1\{1 + f(t)\}$, where f is given, flowing past the fixed boundary $y = 0$ we have, as equation for $u(y, t)$,

$$\frac{\partial u}{\partial t} - V\frac{\partial u}{\partial y} = \frac{\mathrm{d}U}{\mathrm{d}t} + \nu\frac{\partial^2 u}{\partial y^2}. \tag{4.27}$$

The solution of this equation may be written as

$$u = U_1(1 - e^{-Vy/\nu}) + U_1 g(y, t)$$

where g satisfies, from (4.27),

$$\frac{\partial^2 g}{\partial \eta^2} + \frac{\partial g}{\partial \eta} - \frac{\partial g}{\partial \tau} = -\frac{\mathrm{d}f}{\mathrm{d}\tau}, \tag{4.28}$$

where $\eta = Vy/\nu$ and $\tau = V^2 t/\nu$. To solve (4.28), Watson introduces the two-sided Laplace transform

$$\tilde{g}(\eta, s) = s^2 \int_{-\infty}^{\infty} e^{-s\tau} g(\eta, \tau) \, d\tau,$$

so that $\tilde{g}$ satisfies, from (4.28),

$$\frac{\partial^2 \tilde{g}}{\partial \eta^2} + \frac{\partial \tilde{g}}{\partial \eta} - s\tilde{g} = -s\tilde{f}. \tag{4.29}$$

The appropriate solution of equation (4.29) is $\tilde{g} = \tilde{f}(1 - e^{-k\eta})$ where $k = \{1 + (1 + 4s)^{1/2}\}/2$. For a prescribed $f(t)$, $\tilde{f}$ is known, and using the convolution-product rule to find the inverse of $\tilde{g}$ we have, finally,

$$u = U_1 \left[1 - e^{-\eta} + f(\tau) - \frac{\eta \, e^{-\eta/2}}{4\sqrt{\pi}} \int_0^{\infty} \lambda^{-3/2} f(\tau - \lambda) \right.$$
$$\left. \times \exp\left\{ -\left(\lambda + \frac{\eta^2}{16\lambda} \right) \right\} d\lambda \right]. \tag{4.30}$$

From (4.30), with $f(t) = \cos \omega t$, the result of Stuart (1955) set out in equation (4.12) may be recovered. Watson also considers impulsive situations. For example if $U = U_1\{1 + H(\tau)\Delta\}$, where Δ is a constant and $H(\tau)$ the Heaviside unit function, then equation (4.30) yields

$$u = U_1 \left[1 - e^{-\eta} + H(\tau)\Delta \left\{ 1 - \frac{1}{2} e^{-\eta} \mathrm{erfc}\left(\frac{\eta}{4\tau^{1/2}} - \tau^{1/2} \right) \right.\right.$$
$$\left.\left. - \frac{1}{2} \mathrm{erfc}\left(\frac{\eta}{4\tau^{1/2}} + \tau^{1/2} \right) \right\} \right]. \tag{4.31}$$

A multi-step change in free-stream speed, $U = U_1[1 + \{H(\tau) - H(\tau - \tau_0)\}\Delta]$, which reduces it to the original value U_1 after an interval τ_0 is also considered. The shear stress at the boundary, $\tau_w = \mu \partial u/\partial y|_{y=0}$ is given, from (4.31), as

$$\frac{\tau_w}{\rho U_1 V} = 1 + \frac{1}{2} H(\tau)\Delta \left\{ 1 + \mathrm{erf}(\tau^{1/2}) + \frac{e^{-\tau}}{(\pi \tau)^{1/2}} \right\}. \tag{4.32}$$

A further example by Watson has the free stream accelerating, from some initial instant, such that $U = U_1\{1 + H(\tau)\Delta\tau\}$ and the velocity distribution is given from equation (4.30) as

$$u = U_1 \left[1 - e^{-\eta} + \frac{1}{8} H(\tau)\Delta \left\{ 8\tau - (4\tau - \eta) \, \mathrm{erfc}\left(\frac{\eta}{4\tau^{1/2}} - \tau^{1/2} \right) \right.\right.$$
$$\left.\left. - (4\tau + \eta) \, \mathrm{erfc}\left(\frac{\eta}{4\tau^{1/2}} + \tau^{1/2} \right) \right\} \right], \tag{4.33}$$

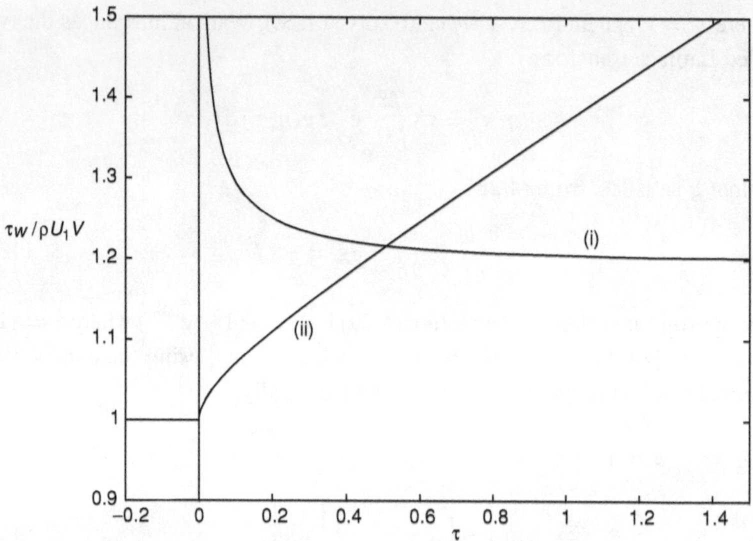

Figure 4.4 The shear stress $\tau_w/\rho U_1 V$ (i) from equation (4.32) following an impulsive change in free-stream speed, and (ii) from equation (4.33) when the free stream accelerates at $t = 0$.

from which the boundary shear stress is given as

$$\frac{\tau_w}{\rho U_1 V} = 1 + \frac{1}{2}H(\tau)\Delta\left\{\tau + \frac{1}{2}\mathrm{erf}(\tau^{1/2}) + \left(\frac{\tau}{\pi}\right)^{1/2}\mathrm{e}^{-\tau} + \tau\,\mathrm{erf}(\tau^{1/2})\right\}.$$

(4.34)

Watson also considers a multiple acceleration that brings the free-stream speed back to its uniform value U_1. In all of these situations, following the change in free-stream conditions, additional vorticity is created at the boundary, initially confined to a layer of thickness $O\{(\nu t)^{1/2}\}$, as in the impulsive flows considered in section 4.2.

The boundary shear stresses for the examples detailed above, given by equations (4.32) and (4.34) are shown in figure 4.4.

As a final example Watson studies the case of a decaying oscillation superposed on the free stream at the initial instant such that $U = U_1\{1 + H(\tau)\Delta\,\mathrm{e}^{-a\tau}\sin b\tau\}$ where $a\ (a > 0)$ and b are constants.

4.4 The angled flat plate

The classical problem of the impulsive motion of an infinite plate, in its own plane, has been treated in section 4.2. The case when the plate is 'bent', so that two semi-infinite planes intersect, along the x-axis, say, at an angle α ($0 < \alpha \leq 2\pi$) has been considered by Sowerby (1951), Hasimoto (1951) and Sowerby and Cooke (1953). The angled plate is set into motion impulsively, at time $t = 0$, with uniform speed U_0 in the x-direction. It proves convenient to introduce cylindrical polar co-ordinates which in this case, for consistency of notation with sections 4.1 and 4.2, are defined such that $z = r \cos \theta$, $y = r \sin \theta$. We assume $\mathbf{v} = \{u(r, \theta, t), 0, 0\}$ so that the continuity equation is satisfied identically and, in the absence of any pressure gradient, the equation satisfied by $u(r, \theta, t)$ may be inferred from equation (1.21) as

$$\frac{\partial u}{\partial t} = \nu \left(\frac{\partial^2 u}{\partial r^2} + \frac{1}{r} \frac{\partial u}{\partial r} + \frac{1}{r^2} \frac{\partial^2 u}{\partial \theta^2} \right). \tag{4.35}$$

The boundary conditions require

$$u \equiv 0, \quad r > 0 \quad \text{at} \quad t = 0,$$

and for $t > 0$,

$$u = U_0 \text{ for } \theta = 0, \alpha, \quad \text{all } r; \quad u \to 0 \text{ as } r \to \infty \text{ for } 0 < \theta < \alpha. \tag{4.36}$$

The most detailed investigation of this problem has been carried out by Sowerby and Cooke who, drawing upon an analogous heat conduction problem, see for example Carslaw and Jaeger (1947), find an infinite series solution for $t > 0$ of (4.35), subject to the boundary conditions (4.36), as

$$u = U_0 \left\{ 1 - \frac{2}{\alpha} \sum_{n=0}^{\infty} \sin s\theta \frac{\Gamma(\frac{1}{2}s)}{\Gamma(s+1)} \eta^s \, e^{-\eta^2} {}_1F_1 \left(\tfrac{1}{2}s + 1; s + 1 : \eta^2 \right) \right\}, \tag{4.37}$$

where $s = (2n + 1)\pi/\alpha$, $\eta = r/2(\nu t)^{1/2}$ and ${}_1F_1$ is the confluent hypergeometric function.

The important special case $\alpha = 2\pi$ was considered earlier by Howarth (1950), and (4.37) recovers his solution as $\alpha \to 2\pi$. In that particular case the configuration is one in which a semi-infinite plate is moved impulsively in its own plane parallel to its edge. Both Sowerby and Cooke, and Howarth, discuss alternative representations of the solution (4.37). However, the series (4.37) converges rapidly close to the intersection $r = 0$ and, in particular, Howarth shows that for his case of $\alpha = 2\pi$ the skin friction, close to the edge, is given by

$$\mu \frac{\partial u}{\partial y} \bigg|_{y=0} \approx -0.46 \mu U_0 (\nu t)^{-3/4} r^{1/2}.$$

The 'edge effect' is shown by Howarth to be confined to a region of scale $O\{(vt)^{1/2}\}$. Beyond this, as $r \to \infty$ along the plate, the classical solution readily emerges.

In his earlier paper Sowerby (1951) had confined attention to the case where $\alpha = \pi/m$, where $m \geq 1$ is an integer, with solution $u_m(r, \theta, t)$. Relatively simple solutions emerge. For example

$$u_1 = U_0\{1 - \mathrm{erf}(\eta \sin \theta)\},$$

$$u_2 = U_0\{1 - \mathrm{erf}(\eta \cos \theta)\mathrm{erf}(\eta \sin \theta)\},$$

$$u_3 = U_0[1 - \mathrm{erf}(\eta \sin \theta) - \mathrm{erf}\{\eta \sin(\pi/3 - \theta)\} + \mathrm{erf}\{\eta \sin(2\pi/3 - \theta)\}],$$

the first of which is, of course, the classical infinite flat plate solution.

In this category of solution we might also mention the work of Levine (1957) who considers the case of an infinite strip, of width $2a$, set into motion with uniform speed U_0.

4.5 Unsteady plate stretching

Devi, Takhar and Nath (1986) have extended the analysis of Wang (1984), for steady three-dimensional flow due to an impermeable stretching plate, to include a particular time dependence. It proves convenient, as in section 2.5.4 to introduce a slight change of notation such that x, y are co-ordinates in the plane, with z perpendicular to it. Then, if the boundary velocities are given by $u = kx(1 - \lambda kt)^{-1}$, $v = ly(1 - \lambda kt)^{-1}$, a self-similar solution is available in which

$$u = kx(1 - \lambda kt)^{-1}f'(\eta), \quad v = ky(1 - \lambda kt)^{-1}g'(\eta),$$
$$w = -(kv)^{1/2}(1 - \lambda kt)^{-1/2}\{f(\eta) + g(\eta)\} \tag{4.38}$$

where $\eta = (k/v)^{1/2}(1 - \lambda kt)^{-1/2}z$. The velocity components in (4.38) satisfy the continuity equation, and equations (1.9), (1.10) then give, as equations for f and g,

$$f''' + (f + g)f'' - f'^2 - \lambda\left(f' + \tfrac{1}{2}\eta f''\right) = 0,$$

$$g''' + (f + g)g'' - g'^2 - \lambda\left(g' + \tfrac{1}{2}\eta g''\right) = 0,$$

together with

$$f(0) = g(0) = 0, \quad f'(0) = 1, \quad g'(0) = r; \quad f', g' \to 0 \quad \text{as} \quad \eta \to \infty,$$

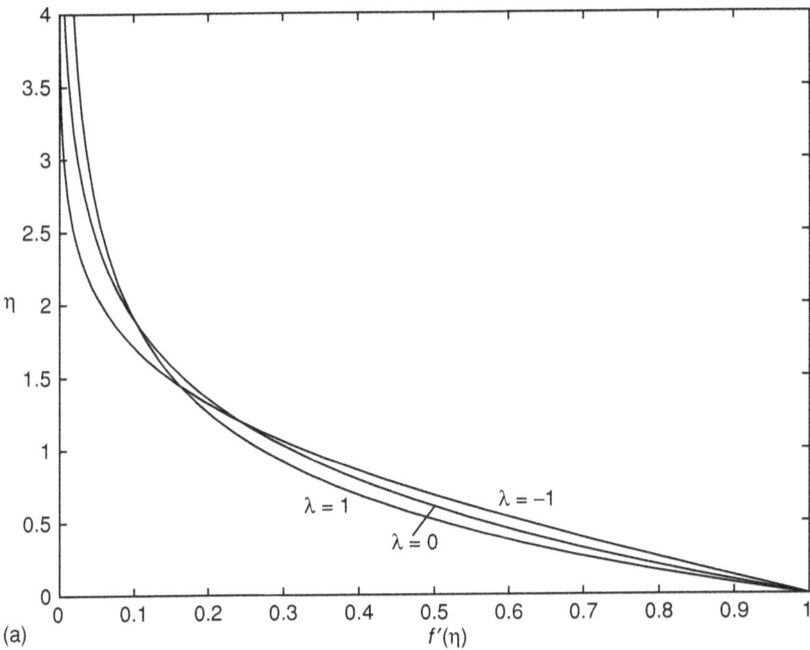

Figure 4.5 (a) Velocity profiles $f'(\eta)$ for $r = \frac{1}{2}$ and $\lambda = -1, 0, 1$. (b) As (a) for the profiles $g'(\eta)$.

where $r = l/k$. This system of equations has been integrated numerically by Devi *et al.* and we present representative velocity profiles in figure 4.5. It may be noted that for the range of decelerating/accelerating flows for which $-1 \leq \lambda \leq 1$ the velocity profiles $f'(\eta)$, $g'(\eta)$ differ little from the steady state stretching solution of Wang (1984).

An interpretation of this solution for $\lambda > 0$ is that of a flow starting from rest at $t = -\infty$ which then accelerates to become unbounded at $t = (\lambda k)^{-1}$ within a layer of vanishing thickness. Devi *et al.* suggest it may find application in extrusion processes.

4.6 Beltrami flows and their generalisation

As we have already noted in section 2.2 Beltrami flows are flows for which $\mathbf{v} \wedge \boldsymbol{\omega} = \mathbf{0}$, and for steady situations, except for the case $\boldsymbol{\omega} \equiv \mathbf{0}$, such flows can only be sustained through the action of a non-conservative body force. For unsteady flow, in the absence of any body force, $\mathbf{v} \wedge \boldsymbol{\omega} = \mathbf{0}$ implies

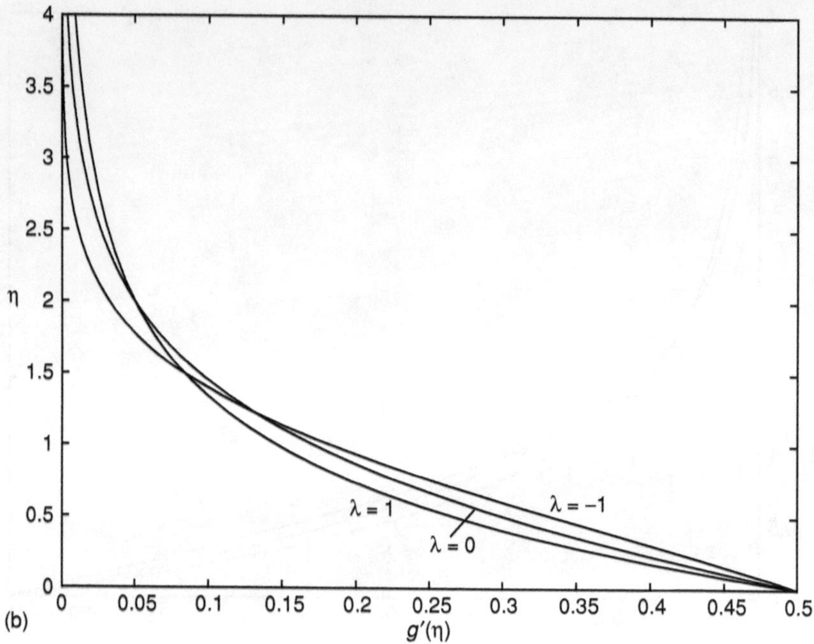

Figure 4.5 (*cont.*)

$$\frac{\partial \boldsymbol{\omega}}{\partial t} = \nu \nabla^2 \boldsymbol{\omega}, \tag{4.39}$$

with

$$\boldsymbol{\omega} = c(\mathbf{x}, t)\mathbf{v}. \tag{4.40}$$

Equation (4.40) shows immediately that such Beltrami flows can be neither planar nor axisymmetric. Setting

$$\mathbf{v} = e^{-c^2 \nu t}\mathbf{f}(\mathbf{x}), \tag{4.41}$$

where c is a constant, Trkal (1919) has shown that both equations (4.39), (4.40) will be satisfied if

$$\nabla \wedge \mathbf{f} = c\mathbf{f}, \tag{4.42}$$

and Berker (1963) has obtained a solution for $\mathbf{f}$ as

$$\mathbf{f} = -\frac{1}{\sqrt{2}}\left(\cos\frac{cx}{\sqrt{2}}\sin\frac{cy}{\sqrt{2}}, \quad -\sin\frac{cx}{\sqrt{2}}\cos\frac{cy}{\sqrt{2}}, \quad -\sqrt{2}\cos\frac{cx}{\sqrt{2}}\cos\frac{cy}{\sqrt{2}}\right). \tag{4.43}$$

As for steady flows considered in section 2.2, a generalisation of these Beltrami flows requires

$$\nabla \wedge (\mathbf{v} \wedge \boldsymbol{\omega}) = \mathbf{0} \tag{4.44}$$

and $\boldsymbol{\omega}$ still satisfying equation (4.39). With $\boldsymbol{\omega} = (0, 0, \zeta)$ and $\mathbf{v} = \nabla \wedge (\psi \mathbf{k})$ as in section 2.2, ζ and the stream function ψ are, from (4.44) again related by equation (2.16), whose solution may now be written as

$$\zeta = -f(\psi, t). \tag{4.45}$$

As before, see equation (1.15), we have

$$\frac{\partial^2 \psi}{\partial x^2} + \frac{\partial^2 \psi}{\partial y^2} = -\zeta \tag{4.46}$$

and equation (4.39) requires

$$\frac{\partial \zeta}{\partial t} = \nu \left(\frac{\partial^2 \zeta}{\partial x^2} + \frac{\partial^2 \zeta}{\partial y^2} \right) \tag{4.47}$$

so that an exact solution of the Navier–Stokes equations results when equations (4.45) to (4.47) are compatible.

A particularly simple form of (4.45) is $\zeta = \alpha \psi$. If we then write $\psi = A \, e^{\beta t} \, \Phi(x, y)$ both equations (4.46), (4.47) will be satisfied provided $\beta = -\nu\alpha$ and Φ satisfies

$$\frac{\partial^2 \Phi}{\partial x^2} + \frac{\partial^2 \Phi}{\partial y^2} + \alpha \Phi = 0.$$

Separation of variables, so that $\Phi(x, y) = F(x)G(y)$, then gives

$$\frac{d^2 F}{dx^2} G + F \frac{d^2 G}{dy^2} + \alpha F G = 0. \tag{4.48}$$

Three special solutions of (4.48) are the following.

(i) $F = \cos ax$, $G = \cos by$. In that case $\alpha = a^2 + b^2$ and

$$\psi = A \cos ax \cos by \, e^{-\nu(a^2+b^2)t}. \tag{4.49}$$

This result was obtained by Taylor (1923) for the special case $a = b$. It may be interpreted as a double array of vortices which decay exponentially with time; the decay rate increases as the scale of the vortices decreases. An example is shown in figure 4.6.

Whilst there is a superficial similarity between (4.41) and (4.49) the essential difference should be noted, namely that in Trkal's solution $\mathbf{v} \wedge \boldsymbol{\omega} = \mathbf{0}$, whilst in Taylor's $\mathbf{v} . \boldsymbol{\omega} = 0$.

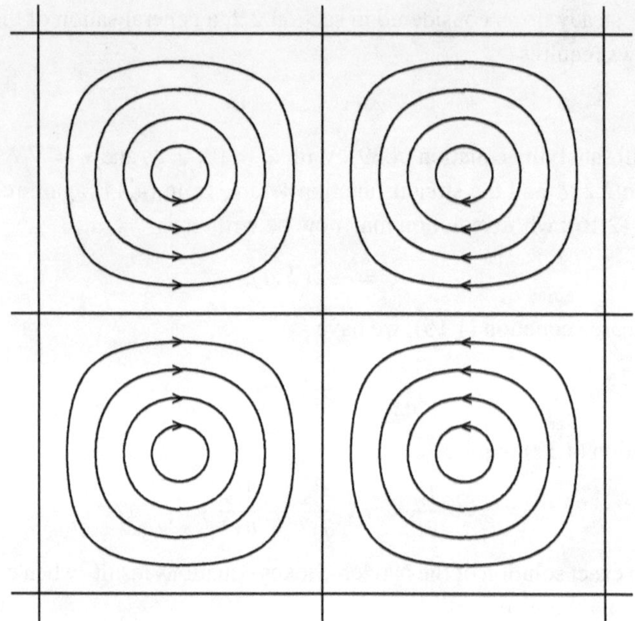

Figure 4.6 Instantaneous streamlines for the double array of vortices considered by Taylor (1923).

(ii) $F = \cosh ax$, $G = \cos by$. For this case $\alpha = b^2 - a^2$ and

$$\psi = A \cosh ax \cos by \, e^{\nu(a^2 - b^2)t}. \tag{4.50}$$

Wang (1966) presents this solution as the viscous analogue of Kelvin's 'cat's eye' vortices, which decay or increase without bound depending upon $a < b$ or $a > b$. A typical streamline pattern is shown in figure 4.7.

(iii) $F = \sinh ax$, $G = \sinh by$. We have $\alpha = -(a^2 + b^2)$ and so

$$\psi = A \sinh ax \sinh by \, e^{\nu(a^2 + b^2)t}. \tag{4.51}$$

Equation (4.51) is a solution that grows without bound, and is interpreted by Wang (1966) as the impingement of two rotational flows. For $|ax|$, $|by| \ll 1$ we have $\psi \propto xy \, e^{\nu(a^2 + b^2)t}$, a time-varying classical stagnation-point flow.

Wang (1989a) generalises the above by writing

$$\zeta = \alpha\{\psi - U(t)y\} \tag{4.52}$$

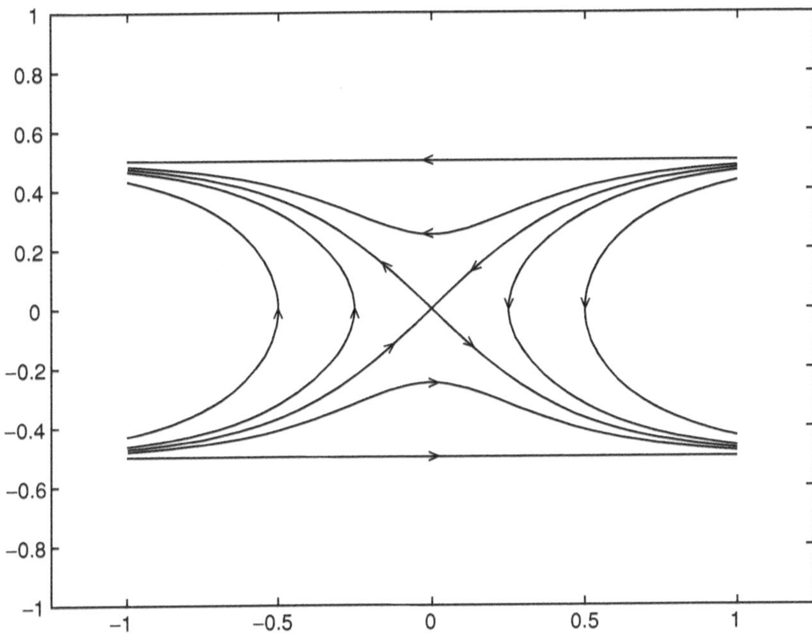

Figure 4.7 An example of the streamline pattern for Wang's (1966) solution, equation (4.50), with $a = b = \pi$ which corresponds, in fact, to a steady state.

so that the vorticity equation (1.17) then yields, as the equation for ψ,

$$\frac{\partial \psi}{\partial t} - \frac{dU}{dt}y + U\frac{\partial \psi}{\partial x} = \nu \nabla^2 \psi. \tag{4.53}$$

If we now write

$$\psi = U(t)y + V(t)F(x, y) \tag{4.54}$$

then from equation (4.46), we have, as the equation for F,

$$\nabla^2 F + \alpha F = 0.$$

We take as the solution of this equation

$$F = B \sin \frac{2\pi y}{h} e^{-(4\pi^2/h^2 - \alpha)^{1/2}x}. \tag{4.55}$$

To complete the solution for ψ in (4.54) we find, from equation (4.53),

$$\frac{dV}{dt} - \left(\frac{4\pi^2}{h^2} - \alpha\right)^{1/2} UV + \alpha\nu V = 0, \tag{4.56}$$

so that V is determined in terms of U as

$$V = \exp\left\{\left(\frac{4\pi^2}{h^2} - \alpha\right)^{1/2}\int U\,\mathrm{d}t - \alpha vt\right\}. \tag{4.57}$$

For the special case $V = U$ the solution of (4.56) is

$$U = \frac{\alpha v}{(4\pi^2/h^2 - \alpha)^{1/2}(1 - e^{\alpha vt})}, \tag{4.58}$$

a result that was anticipated by Berker (1963). If, for example, we now set

$$\alpha = \frac{R(R^2 + 16\pi^2)^{1/2} - R^2}{2h^2} > 0,$$

where $R = U_0 h/v$ is a Reynolds number based on a representative velocity U_0, then from (4.54), (4.55) and (4.58) the solution is

$$\psi = \frac{\alpha v}{(4\pi^2/h^2 - \alpha)^{1/2}(1 - e^{\alpha vt})}\left[y + B\sin\frac{2\pi y}{h}\right.$$

$$\left.\times \exp\left[\left\{R - \sqrt{(R^2 + 16\pi^2)}\right\}x/2h\right]\right]. \tag{4.59}$$

As $t \to -\infty$ this solution, with B arbitrary, approaches the steady solution of Kováasznay, as in equation (2.27), and decays as $t \to \infty$.

Wang (1989a) also considers, with $V \neq U$, a generalisation of his solution (Wang (1966)) of a uniform stream encountering a diverging counterflow which has been outlined in section 2.2.1.

4.7 Stagnation-point flows

In section 2.3 we examined steady stagnation-point flows on an infinite flat plate in different situations, as a model for the flow in the neighbourhood of the stagnation point on a bluff body. Several authors have examined unsteady effects on such flows with a motivation, in part, prompted by aerodynamic flutter problems, and these we now consider.

4.7.1 Transverse oscillations

It is immaterial whether it is the plane boundary or oncoming stream which is oscillating. We assume the former, and that at large distances from the plane the steady two-dimensional flow represented by equation (2.32) is established. This problem has been considered both by Glauert (1956) and Rott (1956), with the boundary $y = 0$ oscillating in its own plane, with frequency ω and velocity amplitude U_0, in the x-direction. In place of (2.32) we now write

$$\psi = (\nu k)^{1/2} x f(\eta) + \left(\frac{\nu}{k}\right)^{1/2} U_0 \, e^{i\omega t} \int_0^\eta g(\xi) \, d\xi \quad \text{where again} \quad \eta = \left(\frac{k}{\nu}\right)^{1/2} y.$$

(4.60)

The pressure is given once more from equation (2.36), so that substitution of (4.60) into equation (1.9) gives

$$k^2 x (f''' + f f'' - f'^2 + 1) + k U_0 \, e^{i\omega t} \left(g'' + f g' - f' g - \frac{i\omega}{k} g \right) = 0,$$

from which we infer that f satisfies (2.35) and g satisfies

$$g'' + f g' - \left(f' + \frac{i\omega}{k} \right) g = 0 \quad \text{with} \quad g(0) = 1, \quad g(\infty) = 0. \quad (4.61)$$

Solutions of equation (4.61) are sought by Glauert and Rott in the form of series, for example

$$g = \sum_{n=0}^{\infty} \left(\frac{i\omega}{k} \right)^n g_n(\eta) \quad (4.62)$$

where the successive terms g_n are determined *seriatim*. The leading term $g_0(\eta)$ corresponds to the case when the plate slides in the x-direction with uniform speed, a case that has been considered in section 2.3. For large values of ω/k the representation (4.62) is not convenient and is replaced by

$$g = \sum_{n=0}^{\infty} \left(\frac{k}{i\omega} \right)^{n/2} G_n(Y), \quad (4.63)$$

where we have $y = (\nu/i\omega)^{1/2} Y$. The leading term now is $G_0 = e^{-Y}$ which is recognised as Stokes' solution (4.7), and indeed with subsequent terms of (4.63) containing a factor e^{-Y} the fluctuating flow is confined to the Stokes layer, thickness $O\{(\nu/\omega)^{1/2}\}$. The skin friction is given by

$$\mu \frac{\partial u}{\partial y} \bigg|_{y=0} = (\rho \mu k)^{1/2} \{ k x f''(0) + U_0 \, e^{i\omega t} g'(0) \}.$$

In figure 4.8 we show $g'(0) = g_r'(0) + i g_i'(0)$, and include results from the series (4.63).

By setting $\omega = -ik$ Rott also considers the case in which the plate slides in the x-direction with speed $U_0 \, e^{kt}$, corresponding to a situation in which the plate accelerates from rest at $t = -\infty$. The solution of (4.61) for g is then, simply, $g = 1 - f'$. Rott also looks at the case when the time dependence is associated with motion in the z-direction. The basic steady stagnation-point

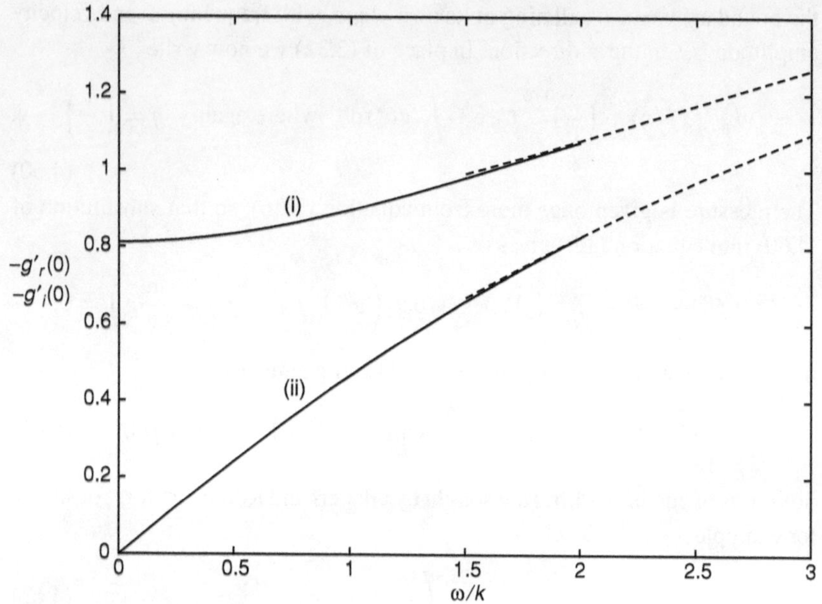

Figure 4.8 The shear stress contributions (i) $g_r'(0)$, (ii) $g_i'(0)$ at a transversely oscillating stagnation point; included in the figure are asymptotic results for $\omega/k \gg 1$, utilising seven terms of the series (4.63).

flow is represented by equation (2.32) with time dependence introduced by setting $w = W_0 h(\eta, \tau)$, with $\tau = kt$ and where h satisfies

$$h'' + fh' - \frac{\partial h}{\partial \tau} = 0, \quad \text{with} \quad h(0, \tau) = 1, \quad h(\infty, \tau) = 0. \tag{4.64}$$

The special case with $h \propto e^{i\tau}$ has been discussed in detail by Rott. Earlier, Wuest (1952), as well as considering more general cases, provided a numerical solution for one particular frequency.

As in section 4.3, Watson (1959) seeks to generalise these unsteady stagnation-point flows, using transform techniques, to accommodate more general plate motions. He writes, instead of (4.60),

$$\psi = (\nu k)^{1/2} x f(\eta) + \left(\frac{\nu}{k}\right)^{1/2} U_0 \int_0^{\eta} g(\xi, \tau) \, d\xi \quad \text{with} \quad \eta \quad \text{as before and} \quad \tau = kt, \tag{4.65}$$

so that g now satisfies

$$g'' + fg' - f'g - \frac{\partial g}{\partial \tau} = 0 \quad \text{with} \quad g(0, \tau) = g_w(\tau) \quad \text{and} \quad g(\infty, \tau) = 0. \tag{4.66}$$

Supposing that the boundary is set into motion at $t = 0$, the Laplace transform

$$\tilde{g}(\eta, s) = \int_0^\infty \mathrm{e}^{-s\tau} g(\eta, \tau)\,\mathrm{d}\tau$$

is introduced, so that $\tilde{g}$ satisfies

$$\tilde{g}'' + f\tilde{g}' - f'\tilde{g} - s\tilde{g} = 0 \quad \text{with} \quad \tilde{g}(0, s) = \tilde{g}_w(s) \quad \text{and} \quad \tilde{g}(\infty, s) = 0. \tag{4.67}$$

It may be noted from (4.61) and (4.67) that the problem for $\tilde{g}/\tilde{g}_w$ is as that considered by Glauert and Rott when s replaces $\mathrm{i}\omega/k$. Watson takes advantage of this by observing that the small-time solution, obtained by expanding $\tilde{g}/\tilde{g}_w$ for large s and inverting term by term can, in principle, be inferred from the series (4.63) and similarly the large-time solution obtained by expanding $\tilde{g}/\tilde{g}_w$ for small s from the series (4.62). In particular for the case of a boundary set into motion with uniform speed U_0 at time $t = 0$ Watson finds, for the skin friction,

$$\mu \left.\frac{\partial u}{\partial y}\right|_{y=0} = (\rho\mu k)^{1/2} \left\{ kx f''(0) + U_0 \left.\frac{\partial g}{\partial \eta}\right|_{\eta=0} \right\}$$

where, from the small-time result

$$\left.\frac{\partial g}{\partial \eta}\right|_{\eta=0} = \frac{a_0}{(\pi\tau)^{1/2}} + 2a_1 \left(\frac{\tau}{\pi}\right)^{1/2} + \sum_{n=0}^\infty \frac{a_{n+3}\tau^{1+n/2}}{\Gamma(2 + \frac{1}{2}n)}, \tag{4.68}$$

where $a_i = \mathrm{d}G_i/\mathrm{d}Y|_{Y=0}$ with $G_i(Y)$ as in (4.63). For large times the solution approaches the steady-state solution of Rott (1956) discussed in section 2.3 for which we have $\lim_{\tau\to\infty} \partial g/\partial\eta|_{\eta=0} = -0.8113$. In figure 4.9 $\partial g/\partial\eta|_{\eta=0}$ is shown, from the series (4.68), and this figure demonstrates how rapidly the steady state solution is reached following the impulsive start.

4.7.2 Orthogonal oscillations

In contrast to the flow in section 4.7.1 above we now consider the situation in which the steady stagnation-point flow is modified by the infinite plane performing harmonic fluctuations in its position along a normal direction. It is convenient, without loss of generality, to deal with the case in which the boundary is fixed with the far-field stagnation-point flow modulated by harmonic fluctuations of arbitrary amplitude and frequency. The study finds application to describe the local dynamics around a stagnation point on an oscillating body, or to model the local effects of disturbances at the stagnation point of a translating bluff body. The problem has received the attention of Grosch and Salwen (1982), Riley and Vasantha (1988), Merchant and Davis (1989) and Blyth and

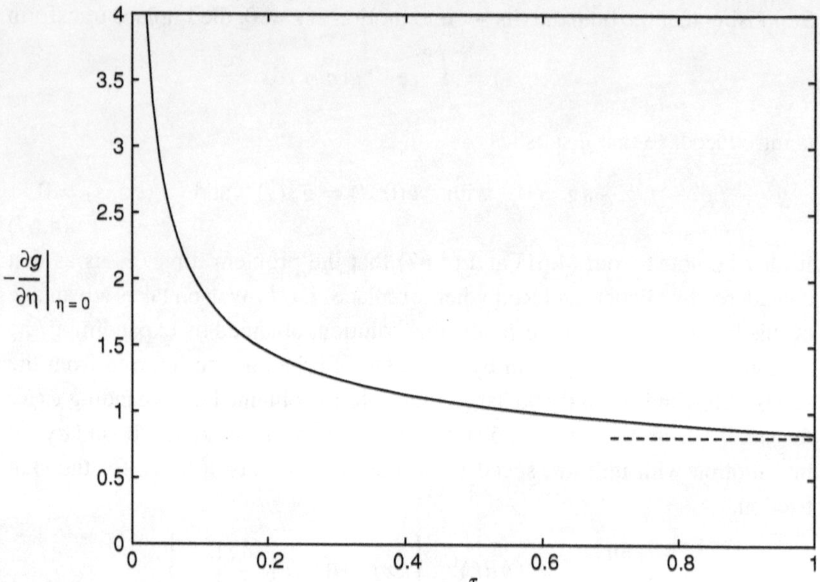

Figure 4.9 The shear stress contribution $\partial g/\partial \eta|_{\eta=0}$ calculated from six terms of the series (4.68).

Hall (2003), where the most recent of these investigations is the most comprehensive.

Far from the plane boundary $y = 0$ the modulated stagnation-point flow has the form

$$u = kx(1 + \Delta \cos \omega t), \quad v = -ky(1 + \Delta \cos \omega t),$$

and so we write, in place of (4.60),

$$\psi = (\nu k)^{1/2} x f(\eta, \tau) \quad \text{where} \quad \eta = \left(\frac{k}{\nu}\right)^{1/2} y \quad \text{and} \quad \tau = \omega t.$$

With the pressure now given by

$$\frac{p_0 - p}{\rho} = \frac{1}{2} k^2 x^2 a^2(\tau) + \frac{1}{2}\omega k x^2 a'(\tau) + \frac{1}{2}\nu k f^2 + \nu k f_\eta - \omega \nu \int_0^\eta f_\tau \, d\eta,$$

$$(4.69)$$

$a(\tau) = 1 + \Delta \cos \tau$, equation (1.9) gives, for f,

$$\sigma f_{\eta\tau} = f_{\eta\eta\eta} + f f_{\eta\eta} - f_\eta^2 + \sigma a' + a^2, \tag{4.70}$$

together with

$$f(0, \tau) = f_\eta(0, \tau) = 0; \quad f_\eta(\infty, \tau) = a(\tau). \tag{4.71}$$

It may be noted that $\sigma = \omega/k$ is the Strouhal number.

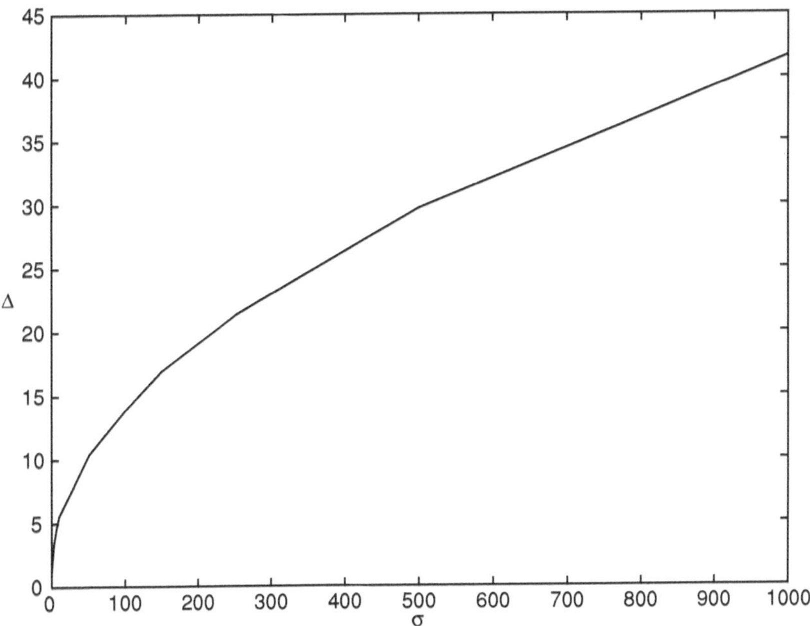

Figure 4.10 The 'barrier', calculated by Blyth and Hall (2003), above which the solution develops a singularity at a finite time.

Grosch and Salwen addressed the problem posed by equations (4.70) and (4.71) for small Δ by expanding the solution as Fourier series in τ with coefficients functions of η, each expanded in powers of Δ. For $\sigma \ll 1$ the solution is a quasi-steady version of the classical Hiemenz solution. For $\sigma \gg 1$ the solution exhibits a double structure. Merchant and Davis take advantage of this double structure for large σ to develop solutions of (4.70), (4.71) when $\Delta \gg 1$. A steady streaming is induced due to the action of Reynolds stresses within the inner Stokes shear-wave layer, thickness $O\{(\nu/\omega)^{1/2}\}$, which persists beyond it, where it is determined as the solution of an ordinary differential equation. Merchant and Davis show that this equation has no solution when $\Delta > 1.289\sigma^{1/2}$.

Blyth and Hall have set these earlier results into a wider context by integrating the differential equation (4.70) forward in time starting from a state of rest. They show that for each value of σ the solution develops a singularity at a finite time for Δ sufficiently large. From their extensive calculations they are able to determine the critical curve in (σ, Δ) parameter space that delineates the regular periodic solutions from those which blow up in a finite time, τ_s, and this result is shown in figure 4.10. In particular as $\sigma \to \infty$ they are able to refine the result of Merchant and Davis, and find $\Delta \sim 1.289\sigma^{1/2} + 0.76$ which

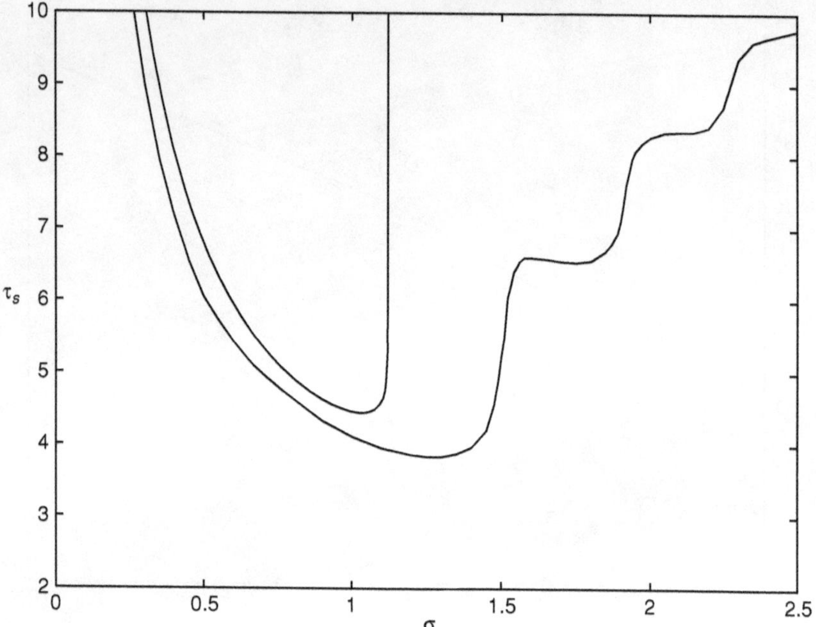

Figure 4.11 Examples showing the 'blow-up' time for $\Delta = 2.0$, Blyth and Hall (2003), and Δ indefinitely large, Riley and Vasantha (1988). In the former case the barrier of figure 4.10 is encountered at $\sigma \approx 1.12$ beyond which the solution is regular.

is almost indistinguishable from the result shown in figure 4.10 for $\sigma > 20$. As $\sigma \to 0$ the critical value of $\Delta \to 1+$. This is a singular limit since for $\sigma \equiv 0$ the steady Hiemenz solution is recovered. The result as $\sigma \to 0$ shows that only the smallest amount of flow reversal in the onset flow is required to initiate a catastrophic failure of the solution.

Riley and Vasantha (1988) were concerned with the case in which the onset flow has zero mean, which in the present context corresponds, formally, to $\Delta \to \infty$. Their results, consistent with those described above, show a breakdown of the solution at a finite time for all non-zero values of σ. In figure 4.11 we compare the breakdown time determined by Blyth and Hall for $\Delta = 2$ with that of Riley and Vasantha.

The failure, or blow-up, of the solution at a finite time is not localised in space but occurs over the entire flow domain with the velocity components $u \propto (\tau_s - \tau)^{-3/2}$, $v \propto (\tau_s - \tau)^{-1}$. In this sense it differs from the problem of blow-up identified by Leray (1934), in which velocities scale with $(\tau_s - \tau)^{-1/2}$ and the singularity is located at a point about which length scales decrease as $(\tau_s - \tau)^{1/2}$. However, in common with Leray the solution discussed here does not have finite total energy, and caution must be exercised before any conclusions are reached about the adequacy, or otherwise, of the Navier–Stokes equations as a mathematical model for incompressible fluid flow in any realistic situation. Vasantha and Riley interpret the singular behaviour as $\tau \to \tau_s-$ as a disruptive event in which a net drift of fluid particles towards the stagnation point leads to an accumulation and consequent eruption of fluid from the stagnation point. In a different, but not dissimilar, situation experiments by Wybrow and Riley (1995) visualise such an eruption as a jet headed by a vortex pair.

4.7.3 Superposed shear flows

The stagnation-point flow, with $\mathbf{v} = (cx, -cy, 0)$, is an exact solution of the Navier–Stokes equations but it cannot satisfy all the usual conditions at any solid boundary. However, in an unbounded region the stresses are everwhere continuous, and the exact solution represents the impingement of two stagnation-point flows, in $y > 0$ and $y < 0$, at the interface $y = 0$. Kambe (1983) superposes on this a shear flow in the x-direction, and simultaneously modulates the basic flow with time by writing

$$u = \frac{\mathrm{d}k}{\mathrm{d}t}x - F(y, t), \qquad v = -\frac{\mathrm{d}k}{\mathrm{d}t}y,$$

where $k = k(t)$. The only non-zero component of vorticity is in the z-direction and has $\zeta = \partial F/\partial y$. From equations (1.13) and (1.16) it satisfies the equation

$$\frac{\partial \zeta}{\partial t} - \frac{\mathrm{d}k}{\mathrm{d}t}y\frac{\partial \zeta}{\partial y} = \nu\frac{\partial^2 \zeta}{\partial y^2}. \tag{4.72}$$

By writing $\eta = e^k y$ and $\tau = \int_0^t e^{2k}\,\mathrm{d}t$, the vorticity transport equation (4.72) transforms to

$$\frac{\partial \zeta}{\partial \tau} = \nu\frac{\partial^2 \zeta}{\partial \eta^2}. \tag{4.73}$$

If we assume $k(0) = 0$, then an initial distribution of vorticity $\zeta(y, 0) = \zeta_0(y) = \zeta_0(\eta)$ leads to a solution of equation (4.73), analogous to that discussed in section 4.2.3, as

$$\zeta(\eta, \tau) = \frac{1}{2}(\pi\nu\tau)^{1/2}\int_{-\infty}^{\infty} \zeta_0(s)\exp\left\{-\frac{(s-\eta)^2}{4\nu\tau}\right\}\,\mathrm{d}s. \tag{4.74}$$

As one example Kambe considers a vortex sheet along the interface $y = 0$ at the initial instant such that

$$F(y, 0) = \begin{cases} U_0, & y > 0, \\ -U_0, & y < 0. \end{cases}$$

This is consistent with $\zeta_0(y) = 2U_0\delta(y)$, where δ is the Dirac delta function. The solution (4.74) is then the Gaussian shear layer

$$\zeta(\eta, \tau) = \frac{U_0}{(\pi \nu \tau)^{1/2}} e^{-(y/a)^2} \quad \text{where} \quad a = 2(\nu\tau)^{1/2} e^{-k}. \tag{4.75}$$

With $\partial F/\partial y = \zeta$ we have, for the velocity field,

$$u = \frac{dk}{dt} x - U_0 e^{-a} \operatorname{erf}(y/a), \quad v = -\frac{dk}{dt} y.$$

For the special case $dk/dt = \text{constant} = c > 0$, $k = ct$ and $\tau = (e^{2ct} - 1)/2c$ so that, as $t \to \infty$, we have from (4.75)

$$\zeta \sim \left(\frac{2c}{\pi \nu}\right)^{1/2} U_0 e^{-ct} e^{-cy^2/2\nu}$$

which implies that, in the final stages, there is a region of vorticity of unchanging shape which decays exponentially with time.

Other examples analysed by Kambe include the case of two vortex sheets which form, initially, a 'top hat' jet for $|y| < y_0$ and the situation in which at the initial instant the distribution of vorticity is represented by two Gaussian distributions centred on $y = -y_0$ and $y = y_0$. Kambe (1986) has also generalised these ideas to three-dimensional situations.

4.7.4 Three-dimensional stagnation-point flow

As in section 2.5.4 we introduce a three-dimensional stagnation-point flow which, as in the steady case, embraces both the two-dimensional and axisymmetric flows. Such a flow has been considered by Cheng, Özişik and Williams (1971). As for the steady case it is convenient to adopt a slight change of notation with x, y as co-ordinates in the plane and z perpendicular to it. Cheng *et al.* assume that far from the boundary the x and y components of velocity are given by $u = kx/(1 + \alpha\omega t), \quad v = ly/(1 + \alpha\omega t)$ where k, l, α and ω are constants, and introduce a self-similar solution as

$$u = \frac{kx}{1 + \alpha\tau} f'(\eta), \quad v = \frac{ky}{1 + \alpha\tau} g'(\eta), \quad w = -\left(\frac{\nu k}{1 + \alpha\tau}\right)^{1/2} \{f(\eta) + g(\eta)\},$$
$$\tag{4.76}$$

where

$$\eta = \left\{\frac{k}{v(1+\alpha\tau)}\right\}^{1/2} z \quad \text{and} \quad \tau = \omega t.$$

With $r = l/k$ and $\sigma = \alpha\omega/k$, introducing (4.76) into equations (1.9) to (1.11) yields the equations for f and g as

$$f''' + (f+g)f'' - f'^2 + 1 + \sigma\left(f' + \tfrac{1}{2}\eta f'' - 1\right) = 0$$
$$\text{with} \quad f(0) = f'(0) = 0, \quad f'(\infty) = 1,$$

$$g''' + (f+g)g'' - g'^2 + r^2 + \sigma\left(g' + \tfrac{1}{2}\eta g'' - r\right) = 0$$
$$\text{with} \quad g(0) = g'(0) = 0, \quad g'(\infty) = r,$$

and the pressure as

$$\frac{p_0 - p}{\rho} = \frac{1}{2(1+\alpha\tau)^2}\{k^2(1-\sigma)x^2 + l^2(1-\sigma)y^2$$
$$+ \frac{kv}{2(1+\alpha\tau)}\{2(f'+g') + (f+g)^2 + \sigma\eta(f+g)\}.$$

Numerical solutions of the equations for f and g are presented by Cheng *et al.* for a range of values of r and σ, although the analogues of the dual solutions of Davey and Schofield (1967) and Schofield and Davey (1967) do not appear to have been explored.

As for the steady case, values of $r > 0$ and $r < 0$ correspond, respectively, to nodal and saddle points of attachment; values of $\alpha > 0$ and $\alpha < 0$ correspond to decelerating and accelerating flows respectively. In common with other self-similar solutions it is not possible to prescribe the velocity profiles (u, v) at some initial instant $t = 0$. Accepting they are given by (4.76) then, for $\alpha > 0$, the flow decays, in a region that increases in thickness $\propto (1 + \alpha\tau)^{1/2}$. By contrast, for $\alpha < 0$ the flow accelerates, in a region of diminishing thickness, and terminates in a singularity at time $t = t_s = (\alpha\omega)^{-1}$. Unlike the situation discussed in the previous section, where the singular behaviour develops spontaneously, the singular behaviour is built in to the solution (4.76).

For $\sigma = 0$ the solution (4.76) coincides with the steady solution (2.77), whilst for $r = 0$ the two-dimensional unsteady solution, first considered by Yang (1958), is recovered. The case $r = 0$, and the axisymmetric flow case $r = 1$, have been considered by Williams (1968). Yang in particular shows that as σ increases, that is as the deceleration rate increases, a value is reached, namely $\sigma = 3.1625$, at which $f''(0) = 0$; for higher deceleration rates there is

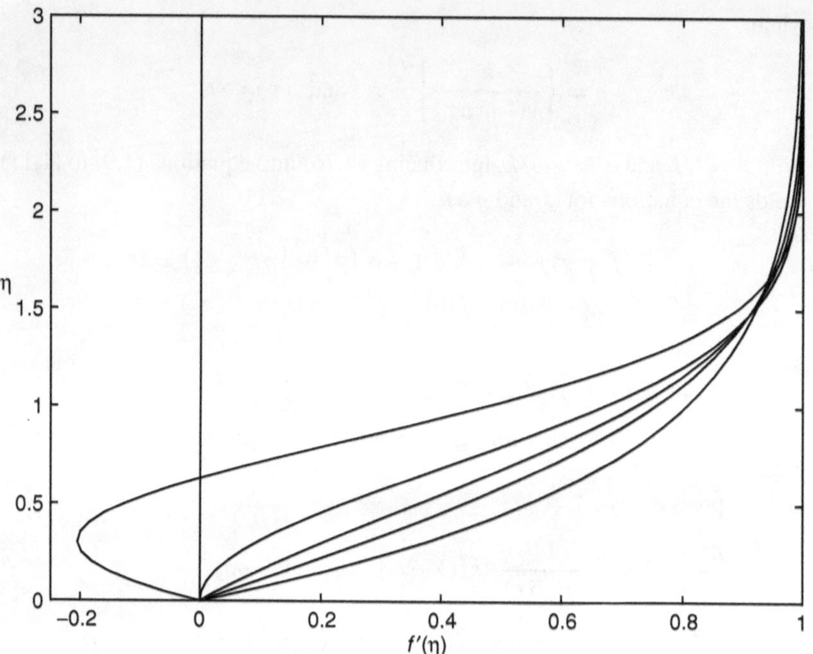

Figure 4.12 Velocity profiles $f'(\eta)$, for the two-dimensional flow corresponding to $r = 0$, for $\sigma = 4.5$ (uppermost), 3.1625, 2.0, 1.0, -1.0.

a region of reversed flow close to the boundary. The corresponding value for $r = 1$ is given by Cheng *et al.* as $\sigma = 3.7173$, at which $f''(0) = g''(0) = 0$. From their results Cheng *et al.* note that for $|\sigma| < 1$ and $0 < r < 1$ the velocity profiles $f'(\eta)$, $g'(\eta)$ differ little from the corresponding steady state profiles. Rajappa (1979) extended Yang's solution to include the effects of hard blowing from the boundary.

In figure 4.12 we show representative velocity profiles for the case of two-dimensional flow $r = 0$.

4.7.5 Rotational three-dimensional stagnation-point flow

Hersh (1970) considers a stagnation-point flow in which there is an unsteady element of lateral vorticity. He takes

$$\mathbf{v} = k\{F(z, t) + (1 - \alpha_0)x, -G(z, t) + (1 + \alpha_0)y, -2z\}, \quad 0 \le \alpha_0 < 1,$$

$$(4.77)$$

for which the corresponding vorticity field is given by $\omega = k(\partial G/\partial z, \partial F/\partial z, 0)$. A self-similar solution is obtained for ω by writing, for the x-component,

$$k\frac{\partial G}{\partial z} = g(\tau)f(\eta) \quad \text{where} \quad \tau = kt \quad \text{and} \quad \eta = \left(\frac{k}{\nu}\right)^{1/2} h(\tau)z. \qquad (4.78)$$

Introducing (4.77) and (4.78) into the vorticity transport equation (1.16) gives

$$\frac{d^2 f}{d\eta^2} + \left(\frac{2h - dh/d\tau}{h^3}\right)\eta\frac{df}{d\eta} + \left\{\frac{(1-\alpha_0)g - dg/d\tau}{gh^2}\right\} f = 0. \qquad (4.79)$$

Equation (4.79) is satisfied provided that

$$h(\tau) = (1 + a\,e^{-4\tau})^{-1/2}, \quad g(\tau) = (1 + a\,e^{-4\tau})^{(\alpha_0 - 1)/4}, \qquad (4.80)$$

where a is an arbitrary constant and, with $\xi = -\eta^2$, f satisfies the confluent hypergeometric equation

$$\xi\frac{d^2 f}{d\xi^2} + \left(\frac{1}{2} - \xi\right)\frac{df}{d\xi} - \frac{(1-\alpha_0)}{4}f = 0.$$

Finally then, from (4.80) and (4.78), the x-component of vorticity is given by

$$k\frac{\partial G}{\partial z} = (1 + a\,e^{-4kt})^{(\alpha_0 - 1)/4} \exp\left\{-\frac{kz^2}{\nu(1 + a\,e^{-4kt})}\right\}$$

$$\times \left\{A\,{}_1F_1\left(\frac{1+\alpha_0}{4}; \frac{1}{2}; \frac{kz^2}{\nu(1 + a\,e^{-4kt})}\right) + B\frac{(k/\nu)^{1/2}z}{(1 + a\,e^{-4kt})^{1/2}}\right.$$

$$\left.\times\,{}_1F_1\left(\frac{3+\alpha_0}{4}; \frac{3}{2}; \frac{kz^2}{\nu(1 + a\,e^{-4kt})}\right)\right\}, \qquad (4.81)$$

where A and B are arbitrary constants. The solution for the y-component of vorticity $k\partial F/\partial z$ follows in a similar manner; the result is as in equation (4.81) with the sign of α_0 changed.

The solution so obtained, with the velocity field following from integration with respect to z, is an exact solution of the vorticity transport equation, but is of limited value. For example, for no choice of A and B can the no-slip condition be satisfied; and the velocity field will only be bounded, as $z \to \infty$, if

$$\frac{A}{B} = -\frac{\Gamma\left(\frac{\alpha_0+1}{4}\right)}{2\Gamma\left(\frac{\alpha_0+3}{4}\right)}.$$

However, the plane $z = 0$ can only represent the interface between two opposing streams if $A \equiv 0$, otherwise the shear stress is discontinuous there.

4.7.6 Flow at a rear stagnation point

In section 2.3 we considered the classical two-dimensional steady stagnation-point flow on the plane boundary $y = 0$ due to Hiemenz (1911). This was proposed as the flow that prevails at the front stagnation point of a blunt cylindrical body in steady flow. At a rear stagnation point, on a circular cylinder say, the situation is different. Common observation shows that following an impulsive start when the flow is everywhere irrotational, with the exception of a vortex sheet at the surface, reversed flow develops as the region of vortical flow thickens rapidly. In contrast with the front stagnation point, where a balance is achieved between diffusion of vorticity and the inhibiting onset flow resulting in a steady state, diffusion of vorticity is reinforced by advection to ensure that no steady state is possible. Proudman and Johnson (1962) model this by considering a 'rear' stagnation-point flow on a plane boundary. For an inviscid fluid the appropriate irrotational flow is simply $u = -kx$, $v = ky$. For a viscous fluid it is then natural to write, instead of (2.32),

$$\psi = -(\nu k)^{1/2} x f(\eta, \tau) \quad \text{where} \quad \tau = kt \quad \text{and} \quad \eta = \left(\frac{k}{\nu}\right)^{1/2} y, \qquad (4.82)$$

and for the pressure, instead of (2.36), we have

$$\frac{p_0 - p}{\rho} = \frac{1}{2} k^2 x^2 + \nu k \left(\frac{1}{2} f^2 + f_\eta + \int_0^\eta f_\tau \, d\eta\right).$$

Equation (1.9) then gives, for $f(\eta, \tau)$,

$$f_{\eta\tau} - f_\eta^2 + f f_{\eta\eta} - f_{\eta\eta\eta} = -1,$$
$$\text{with} \quad f(0, \tau) = f_\eta(0, \tau) = 0, \quad f_\eta(\infty, \tau) = 1. \qquad (4.83)$$

Equation (4.83) may be compared with (2.35). In particular it should be noted that if a steady state is assumed such that $f_{\eta\tau} \equiv 0$, the resulting equation has no solution.

Proudman and Johnson attack equation (4.83) by a variety of analytical and numerical methods, with the vortex sheet at the surface as an initial state. Their investigation has been extended by Robins and Howarth (1972). Principal amongst the results obtained are (a) that flow reversal first takes place close to $y = 0$ at $\tau \approx 0.65$, and that thereafter the flow divides into a wall region of reversed flow and an outer region of forward flow as required by the condition at infinity in (4.83), and (b) that as $\tau \to \infty$ the vortical layer thickness $\delta_v = O\{(\nu/k)^{1/2} e^\tau\}$. This latter result demonstrates that the dynamics of the flow close to the rear stagnation point with its flow reversal must not be confused with onset of boundary-layer separation. At any finite time, as $\nu \to 0$, $\delta_v \to 0$, whereas boundary-layer separation is associated with an unbounded δ_v at a finite time as $\nu \to 0$. The apparent paradox was resolved by van Dommelen and Shen (1980). Their investigation shows that for a circular cylinder the solution of the boundary-layer equations breaks down at a finite time, namely $\tau_s \approx 1.5$, away from the rear stagnation point.

4.8 Channel flows

As for steady flows, unsteady flow in parallel-sided channels has attracted attention with and without transpiration at the two boundaries. Such investigations may find application in the modelling of haemodialysis, pulsating diaphragms, transpiration cooling, isotope separation and filtration. An obvious source of unsteadiness is a time-dependent pressure gradient, but unsteadiness due to movement of the bounding channel walls either in their own planes, or more interestingly normal to them, is also possible.

4.8.1 Fixed boundaries

Wang (1971) has considered the case of a channel with fixed, but porous, walls distance h apart, across which fluid is drawn with uniform speed V and between which there is a fluctuating pressure gradient

$$\frac{1}{\rho}\frac{\partial p}{\partial x} = P_1 + P_2 \cos \omega t.$$

With $\mathbf{v} = \{u(y, t), V, 0\}$ the x-component of velocity may be separated into a time-dependent and time-independent part by writing $u(y, t) = u_1(y) + u_2(y)\,e^{i\omega t}$, where the real part is to be understood. The solution for u_1 is determined, from equation (2.3), as

$$\frac{u_1}{P_1 h/V} = \frac{e^{R\eta} - 1}{e^R - 1} - \eta,$$

where $R = Vh/\nu$ and $\eta = y/h$. For the time-dependent part of u, which from equation (1.9) satisfies

$$\frac{\partial u_2}{\partial t} + V\frac{\partial u_2}{\partial y} = -P_2 e^{i\omega t} + \nu\frac{\partial^2 u_2}{\partial y^2}, \tag{4.84}$$

we may write $u_2(y, t) = (P_2 h^2/\nu)f(\eta)\,e^{i\omega t}$, so that from (4.84) f satisfies

$$f'' - Rf' - iM^2 f = 1, \quad \text{where} \quad M^2 = \omega h^2/\nu. \tag{4.85}$$

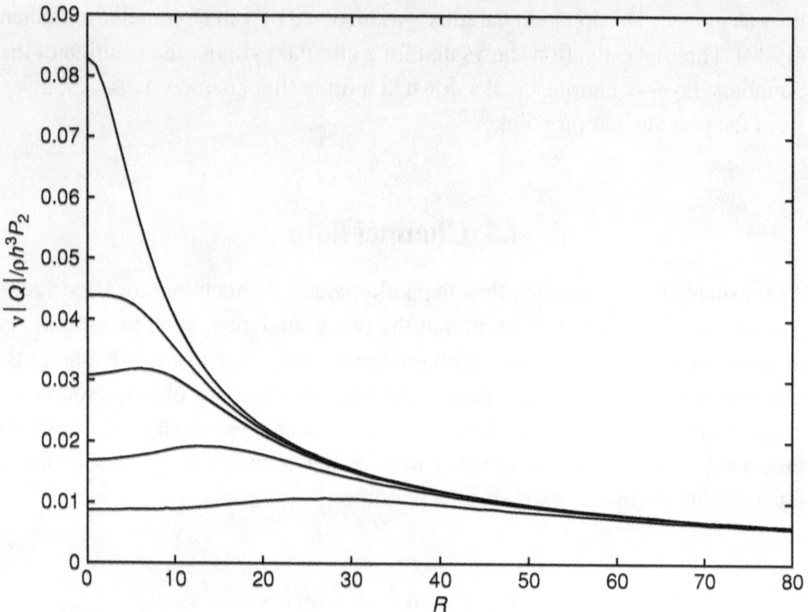

Figure 4.13 The magnitude of the instantaneous mass flux along the channel for cases $M = 1$ (uppermost), 4, 5, 7, 10.

And with $f(0) = f(1) = 0$, the required solution of equation (4.85) is

$$\frac{u_2}{P_2 h^2/\nu} = \frac{i}{M^2}\left\{1 + \frac{(1 - e^{\lambda_2})e^{\lambda_1 \eta} - (1 - e^{\lambda_1})e^{\lambda_2 \eta}}{e^{\lambda_2} - e^{\lambda_1}}\right\} e^{i\omega t}, \qquad (4.86)$$

with

$$\lambda_{1,2} = \frac{R \pm (R^2 + 4iM^2)^{1/2}}{2}.$$

For small R both u_1 and u_2 are essentially symmetric about the channel centre-line $y = h/2$, regardless of the value of M. As R increases both exhibit a degree of asymmetry which is reflected in the mass flux along the channel. The instantaneous mass flux Q is given by, using (4.86),

$$\frac{Q}{\rho h^3 P_2/\nu} = \frac{i\,e^{i\omega t}}{M^2}\left\{1 + \frac{(e^{\lambda_1} - 1)(e^{\lambda_2} - 1)(\lambda_1 - \lambda_2)}{\lambda_1 \lambda_2 (e^{\lambda_2} - e^{\lambda_1})}\right\},$$

whose magnitude is shown as a function of R in figure 4.13 for various values of M.

4.8.2 Squeeze flows

The squeezing of a fluid between two plates may be considered as a model for the unsteady loading of mechanical parts as, for example, in thrust bearings. For such a configuration the stagnation-point flow considered in section 4.7.4 suggests a self-similar form of solution in which for plates situated at $y = y_p = \pm h(t)$ we have $h(t) = h_0(1 - \alpha t)^{1/2}$. This is the approach adopted by Wang (1976) who takes

$$\psi = \frac{\alpha h_0 x}{2(1 - \alpha t)^{1/2}} f(\eta) \quad \text{where} \quad \eta = \frac{y}{h} = \frac{y}{h_0(1 - \alpha t)^{1/2}},$$

so that

$$u = \frac{\alpha x}{2(1 - \alpha t)} f'(\eta) \quad \text{and} \quad v = -\frac{\alpha h_0}{2(1 - \alpha t)^{1/2}} f(\eta), \tag{4.87}$$

and the equation for $f(\eta)$ is, from (1.9),

$$\frac{2 \operatorname{sgn} \alpha}{M^2} f''' + ff'' - f'^2 - \eta f'' - 2f' = A, \tag{4.88}$$

where $M^2 = |\alpha| h_0^2 / \nu$ and A is a constant. Since the flow is symmetrical about $y = 0$ equation (4.88) is to be solved subject to the boundary conditions

$$f(0) = f''(0) = 0; \quad f(1) = 1, \quad f'(1) = 0, \tag{4.89}$$

and the constant A then takes the value $A = 2 \operatorname{sgn} \alpha f'''(1)/M^2$, which is not determined *a priori*. The pressure is now given by

$$\frac{p_0 - p}{\rho} = -\frac{A \alpha^2 x^2}{8(1 - \alpha t)^2} + \frac{\alpha^2 h_0^2}{8(1 - \alpha t)} \left(f^2 - 2\eta f + \frac{4 \operatorname{sgn} \alpha}{M^2} f' \right),$$

where p_0 may now be a function of t.

For $\alpha > 0$ the plates draw together and touch at $t = \alpha^{-1}$, when all other flow variables become unbounded. The velocity profiles $f'(\eta)$ for this case are shown in figure 4.14. As M increases , boundary layers develop adjacent to the plates, and the bulk of the fluid is squeezed out with uniform velocity. For $\alpha < 0$ the plates are drawn apart and, depending upon the value of M, the profiles exhibit different characteristics as shown in figure 4.15. For sufficiently small M fluid is drawn inwards at all points in the gap; but as M increases we see that an outflow develops in the neighbourhood of the boundaries.

There is a qualitative similarity between the velocity profiles shown in figures 4.14 and 4.15 and the steady state situation in which fluid is injected or withdrawn across the fixed porous boundaries of a channel as discussed in chapter 2. In particular, boundary layers form as M increases when the plates are squeezed together, whilst as they are drawn apart a region of reversed flow develops as M increases.

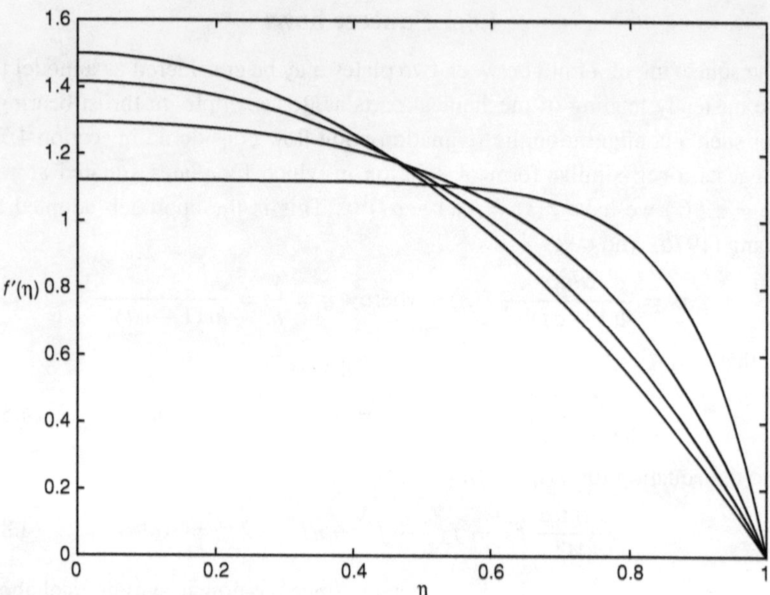

Figure 4.14 Velocity profiles $f'(\eta)$ for squeeze flows corresponding to $\alpha > 0$ with $M = 0.0$ (uppermost), 1.498, 2.879, 7.204.

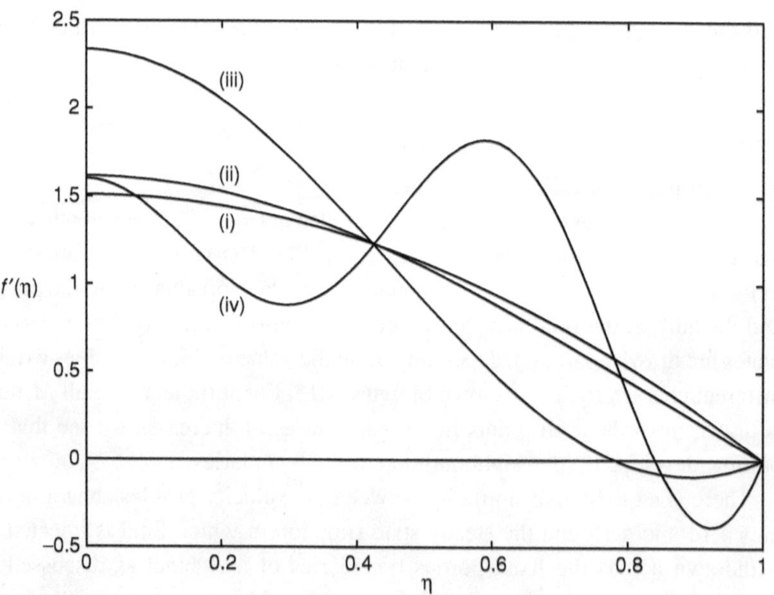

Figure 4.15 Velocity profiles $f'(\eta)$ for squeeze flows corresponding to $\alpha < 0$ with (i) $M = 0.447$, (ii) 1.399, (iii) 2.2214, (iv) 7.195.

Dauenhauer and Majdalani (2003) have extended the range of solutions of equation (4.88) by allowing transpiration across the moving boundaries at $y = \pm h$. The boundary condition on f at $\eta = 1$ in (4.89) is modified to accommodate this. Results are presented for various injection/suction rates with the boundaries closing together or moving apart.

The possibility of multiple solutions, a feature of the steady flow between porous boundaries discussed in section 2.4.1, does not appear to have been addressed in either of the above investigations.

In an earlier paper Thorpe (1967) considered the general case in which the plates are in position $y_p = \pm h(t)$. By writing

$$u = -\frac{1}{h}\frac{dh}{dt}xf'(\eta), \quad v = \frac{dh}{dt}f(\eta) \quad \text{where} \quad \eta = \frac{y}{h}, \tag{4.90}$$

Thorpe shows from equation (1.9) that $\partial p/\partial \eta$ is again independent of x, so that $\partial^2 p/\partial \eta \partial x = 0$. Differentiating (1.9) with respect to y, and u, v as in (4.90) gives

$$\frac{h}{v}\frac{dh}{dt}(2f'' + \eta f''' + f'f'' - ff''') + f^{iv} = \frac{h^2}{vdh/dt}\frac{d^2h}{dt^2}f''. \tag{4.91}$$

At this stage the similarity-solution procedure would seek to ensure that the coefficients of the bracketed expression on the left-hand side of equation (4.91) and of f'' on its right-hand side are constant. Such a procedure leads directly to the form of solution in equation (4.87). However, Thorpe noted that if

$$f = \eta + \frac{(-1)^{n+1}}{n\pi}\sin n\pi\eta, \tag{4.92}$$

then equation (4.91) will be satisfied provided that $h(t)$ satisfies the non-linear equation

$$\frac{d^2h}{dt^2} + (n\pi)^2\frac{v}{h^2}\frac{dh}{dt} - \frac{3}{h}\left(\frac{dh}{dt}\right)^2 = 0.$$

A detailed study of this equation does not appear to have been carried out. However, one solution is readily shown to be $h = h_0(1 - \alpha t)^{1/2}$ where $\alpha = -(n\pi)^2 v/2h_0^2$ which then leads to (4.87), where f satisfies equation (4.88) with $M^2 = (n\pi)^2/2$, whose solution is given by (4.92). This case is included in figure 4.15 with $M = \pi/\sqrt{2} = 2.2214$ corresponding to $n = 1$, and is characterised by the vanishing of f'' at $\eta = 1$.

Hall and Papageorgiu (1999) have considered the case when $h(t)$ is a prescribed periodic function of time. They demonstrate that the flow may break down chaotically following an instability associated with the time dependence. The amplitude of the oscillation that the squeeze flow can support before breakdown is determined; and it is noted that load bearing properties may change significantly when breadown occurs.

4.8.3 Periodic solutions

Watson, Banks, Zaturska and Drazin (1991) and Cox (1991b) have reconsidered the steady channel flows of section 2.4.1, in particular for asymmetric flows. The former achieve this by allowing the boundaries to stretch, that is $u \propto x$ at $y = \pm h$, at different rates, whilst Cox, with both boundaries fixed, has transpiration across one with the other impermeable. In both of these situations it is shown that time-periodic limit-cycle solutions emerge following Hopf bifurcations.

Wave-like flows have been considered by Hui (1987) as follows. With $\alpha = -K$ and with U constant equations (4.52) and (4.53) become

$$\nabla^2 \psi = K(\psi - Uy) \quad \text{and} \quad \frac{\partial \psi}{\partial t} + U \frac{\partial \psi}{\partial x} = \nu \nabla^2 \psi \quad \text{respectively.}$$

Then, letting $\Psi = \psi - Uy$ in the above gives

$$\nabla^2 \Psi = K \Psi \quad \text{and} \quad \frac{\partial \Psi}{\partial t} + U \frac{\partial \Psi}{\partial x} = \nu K \Psi. \tag{4.93}$$

The second of equations (4.93) is satisfied identically by $\Psi = e^{\nu K t} F(X, y)$ where $X = x - Ut$; letting $F(X, y) = f(\xi)$ where $\xi = X \cos \beta + y \sin \beta$, with β constant, the first of (4.93) becomes $f'' - Kf = 0$. With $K = -k^2$, for example, $f(\xi) = A \cos\{k(\xi + B)\}$ for constant A, B so that, finally,

$$\psi = Uy + A\,e^{-\nu k^2 t} \cos\{k(x \cos \beta + y \sin \beta - Ut \cos \beta)\} \tag{4.94}$$

which represents a decaying, travelling, periodic wave superposed on a uniform stream. This plane wave, wavenumber k, propagates with speed $U \cos \beta$ at an angle β to the x-axis. It may be noted that since $U \cos \beta$ is the component of the free-stream velocity in a direction normal to the wave fronts, signals are carried by the undisturbed uniform flow. Streamline patterns from (4.94), at different values of t, are shown in figure 4.16.

Hui discusses another class of solutions by writing $\Psi = e^{\nu K x/U} g(\xi)$, with ξ as before. Again the second of equations (4.93) is satisfied identically whilst the first now requires

$$g'' + \frac{2\nu K}{U} \cos \beta g' + K\left(\frac{\nu^2 K}{U^2} - 1\right) g = 0,$$

so that with $g \propto e^{\lambda \xi/U}$ we have the two possibilities for λ, namely

$$\lambda_{1,2} = -\nu K \cos \beta \pm K \nu \left(\frac{U^2}{K\nu^2} - \sin^2 \beta\right)^{1/2}.$$

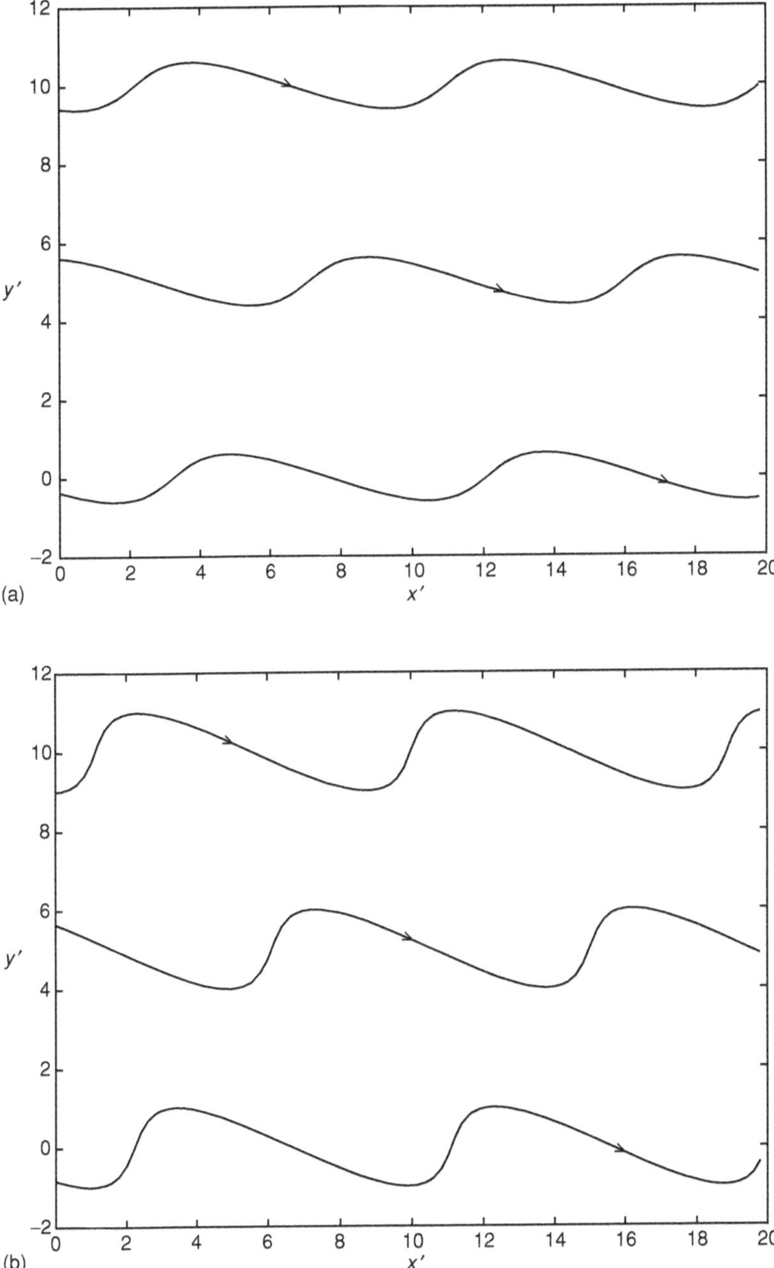

Figure 4.16 Streamline patterns for the wave superposed on a uniform stream, equation (4.94). In these illustrations the variables are dimensionless such that in (4.94) a typical length is taken as k^{-1}, time $(kU)^{-1}$ and a stream function U/k. Values $\beta = \pi/4$, $Ak/U = 1$ and $R = U/\nu k = 2$ have been chosen. For the streamlines $\psi' = 10$ (uppermost), $5, 0$. (a) $t' = 0$. (b) $t' = 1$.

An example, with $K = -k^2$ gives

$$\psi = Uy + \exp\left\{-\frac{vk^2}{U}\left(x \sin^2 \beta - \frac{1}{2}y \sin 2\beta + Ut \cos^2 \beta\right)\right\}$$
$$\times \cos[k\{1 + (vk/U)^2 \sin^2 \beta\}^{1/2}\{x \cos \beta + y \sin \beta - Ut \cos \beta\}],$$

which again represents a periodic travelling wave whose amplitude now varies both spatially and temporally.

Results similar to those obtained by Hui may be inferred from the papers by Grauel and Steeb (1985) and Moore (1991).

Craik and Criminale (1986) also superpose disturbances on basic shear flows, which may be two- or three-dimensional possessing spatially uniform, but not necessarily steady, strain rates so that

$$\mathbf{V} = \mathbf{Sx} + \mathbf{V}_0(t) \quad \text{where} \quad \mathbf{S} \equiv \{\sigma_{ij}(t)\}.$$

The disturbance is denoted by $\mathbf{v}'$ so that $\mathbf{v} = \mathbf{V} + \mathbf{v}'$. Since the 'disturbance' may have arbitrary amplitude the resultant velocity field $\mathbf{v}$ is an exact solution of the Navier–Stokes equations. The disturbance is written as

$$\mathbf{v}' = \hat{\mathbf{v}}(t)F(\eta) \quad \text{where} \quad \eta = \boldsymbol{\alpha}(t).\mathbf{x} + \delta(t), \tag{4.95}$$

and the quantities α and δ are chosen such that

$$\frac{d\boldsymbol{\alpha}}{dt} + \mathbf{S}^T \boldsymbol{\alpha} = \mathbf{0} \quad \text{and} \quad \frac{d\delta}{dt} + \boldsymbol{\alpha}.\mathbf{V}_0 = 0.$$

From (4.95) the continuity equation (1.12) is satisfied provided $\hat{\mathbf{v}}.\boldsymbol{\alpha} = 0$ and making use of that and, furthermore, setting

$$F(\eta) = A \cos K\eta + B \sin K\eta,$$

where A, B, K are constants, Craik and Criminale show that, when the body force is zero, $\hat{\mathbf{v}}$ satisfies

$$\frac{d\hat{\mathbf{v}}}{dt} + \mathbf{S}\hat{\mathbf{v}} = -vK^2\boldsymbol{\alpha}.\boldsymbol{\alpha}\hat{\mathbf{v}} + 2\mathbf{T}\hat{\mathbf{v}}/\boldsymbol{\alpha}.\boldsymbol{\alpha} \tag{4.96}$$

where, with $\boldsymbol{\beta} = \mathbf{S}^T \boldsymbol{\alpha}$, $\mathbf{T} = (\beta_1\boldsymbol{\alpha}, \beta_2\boldsymbol{\alpha}, \beta_3\boldsymbol{\alpha})$.

Although formulated generally, Craik and Criminale consider steady basic flows such that $\mathbf{S}$, $\mathbf{V}_0$ are constant. In particular they concentrate on two-dimensional flows with

$$\mathbf{S} = \begin{pmatrix} 0 & 0 & 0 \\ 0 & a & b \\ 0 & -b & -a \end{pmatrix}, \tag{4.97}$$

and three-dimensional flows with

$$S = \begin{pmatrix} a & 0 & 0 \\ 0 & b & 0 \\ 0 & 0 & -a-b \end{pmatrix}. \tag{4.98}$$

In (4.97) the degenerate case $a = \pm b$ corresponds to the problem addressed by Kelvin (1887) but in a different co-ordinate frame. Kelvin's shear matrix is

$$S = \begin{pmatrix} 0 & 0 & k \\ 0 & 0 & 0 \\ 0 & 0 & 0 \end{pmatrix},$$

for which $\alpha = \{\alpha_1, \alpha_2, \alpha_3(t)\}$ where α_1, α_2 are constants and $\alpha_3(t) = \alpha_3(0) - k\alpha_1 t$. The solution of (4.94) for the third component may then be written as

$$\hat{w}(t) = \exp\left[-\frac{\nu K^2}{3k\alpha_1}\{3k\alpha_1(\alpha_1^2 + \alpha_2^2)t - \alpha_3(t)^2\}\right]/\alpha.\alpha,$$

and $\hat{u}$, $\hat{v}$ follow by direct integration. This particular case was first considered by Kelvin, who was apparently unaware of its status as an exact solution of the Navier–Stokes equations believing that $|\mathbf{v}'| \ll 1$ was a necessary condition for the disturbance.

Craik and Criminale also discuss two-dimensional flows, (4.97), for $a^2 \neq b^2$, $a = 0$ and $b = 0$; and for three-dimensional flows, (4.98), for $a = b, 2a + b = 0$ and $a + 2b = 0$.

5

Unsteady axisymmetric and related flows

5.1 Pipe and cylinder flows

In the first part of this chapter we consider the unsteady flow within circular pipes either driven by a pressure gradient that may have both a steady and time-dependent part, or steady, with unsteadiness of the flow introduced by an unsteady motion of the pipe itself. By cylinder flows we imply the motion external to a pipe, induced by motion of the pipe itself. With the velocity vector $\mathbf{v} = \mathbf{v}(r, t)$ the governing equations are as set out in chapter 3 with the components of $\partial\mathbf{v}/\partial t$ added to equations (3.2) to (3.4) as appropriate.

5.1.1 Impulsive pipe flow

Szymański (1932) considered situations for which the fluid is at rest for $t < 0$, so that with $p = -P(t)z$ we have $P(t) \equiv 0$ for $t < 0$, and for $t > 0$ the flow is unidirectional with $v_z(r, t)$ satisfying

$$\frac{\partial v_z}{\partial t} = \frac{P}{\rho} + v \left(\frac{\partial^2 v_z}{\partial r^2} + \frac{1}{r} \frac{\partial v_z}{\partial r} \right), \tag{5.1}$$

with

$$v_z(r, 0) = 0, \quad 0 \le r \le a; \quad v_z(a, t) = 0, \quad \text{all} \quad t. \tag{5.2}$$

For the particular case $P = \text{constant}$ for $t > 0$, the flow approaches Poiseuille flow as $t \to \infty$ and Szymański writes

$$v_z = \frac{P}{4\mu} \left(a^2 - r^2 \right) + \bar{v}_z(r, t), \tag{5.3}$$

so that with $\bar{v}_z(r, t) = \bar{V}(r)\, e^{-v\lambda^2 t/a^2}$, $\bar{V}(r)$ satisfies

$$r\bar{V}'' + \bar{V}' + (\lambda^2 r/a^2)\bar{V} = 0,$$

128

with solution that is bounded at $r = 0$ and vanishes at $r = a$, $\bar{V}(r) = J_0(\lambda_n r/a)$ where λ_n is a solution of $J_0(\lambda_n) = 0$. The complete solution of (5.3) may then be written as

$$v_z = \frac{P}{4\mu}\left(a^2 - r^2\right) + \sum_{n=0}^{\infty} C_n J_0(\lambda_n r/a)\, \mathrm{e}^{-\nu \lambda_n^2 t/a^2},$$

where the constants C_n are determined from the condition $v_z(r, 0) = 0$ as

$$C_n = \frac{2Pa^2}{\mu \lambda_n^3 J_0'(\lambda_n)}.$$

Szymański also considers cases for which $P(t)$ is a periodic function of time, and $P(t) \to$ constant as $t \to \infty$; whilst Smith (1997) addresses the case for which $P(t) \propto k^2 \mathrm{e}^{-\nu k^2 t}$.

5.1.2 Periodic pipe flow

Situations for which $p = -Pz + f(t)z$ where $f(t)$ is a periodic function of time for all t have been studied by Sexl (1930), Lambossy (1952), Womersley (1955a, 1955b) and Uchida (1956). In Womersley's case the motivation for the study was physiological, in connection with the flow of blood in the larger arteries. Since $f(t)$ can be represented as a Fourier series, and since the governing equations are linear we may, without loss of generality, consider a single frequency and let

$$\frac{\partial p}{\partial z} = -P - c_1 \cos \omega t - c_2 \sin \omega t = -P - c\, \mathrm{e}^{i\omega t}, \quad c = c_1 - ic_2, \quad (5.4)$$

where, as in chapter 4, the real part of any complex quantity is to be understood. We may now write

$$v_z(r, t) = \frac{P}{4\mu}\left(a^2 - r^2\right) + \tilde{v}_z(r, t) \tag{5.5}$$

where, from (5.1) and (5.4), $\tilde{v}(r, t)$ satisfies

$$\frac{\partial \tilde{v}_z}{\partial t} = \frac{c}{\rho}\mathrm{e}^{i\omega t} + \nu \left(\frac{\partial^2 \tilde{v}_z}{\partial r^2} + \frac{1}{r}\frac{\partial \tilde{v}_z}{\partial r}\right) \tag{5.6}$$

and, seeking a separable solution with $\tilde{v}_z = c\tilde{V}(r)\, \mathrm{e}^{i\omega t}/\rho$ we have

$$r\tilde{V}'' + \tilde{V}' + i^3 k^2 r \tilde{V} = -\frac{r}{\nu} \quad \text{where} \quad k^2 = \frac{\omega}{\nu}. \tag{5.7}$$

The solution of (5.7) for $\tilde{V}$ which is bounded at $r = 0$, and satisfies the no-slip condition at $r = a$, is

$$\tilde{V} = \left(\frac{J_0(i^{3/2}kr)}{J_0(i^{3/2}ka)} - 1\right)\frac{i}{\omega},$$

so that if we introduce Kelvin's functions ber and bei with

$$J_0(i^{3/2}x) = \text{ber}(x) + i\text{bei}(x),$$

we have, finally,

$$\tilde{v}_z = \left\{\frac{c_1 F_2 + c_2(F_1 - 1)}{\rho\omega}\right\}\cos\omega t + \left\{\frac{c_2 F_2 + c_1(F_1 - 1)}{\rho\omega}\right\}\sin\omega t,$$

where

$$F_1(kr) = \frac{\text{ber}(kr)\text{ber}(ka) + \text{bei}(kr)\text{bei}(ka)}{\text{ber}^2(ka) + \text{bei}^2(ka)},$$

$$F_2(kr) = \frac{\text{ber}(kr)\text{bei}(ka) - \text{bei}(kr)\text{ber}(ka)}{\text{ber}^2(ka) + \text{bei}^2(ka)}.$$

The instantaneous volume flux across any section is given by

$$Q = \int_0^a 2\pi v_z r \, dr$$

$$= \frac{P\pi a^4}{8\mu} + \frac{\pi a^4}{\mu}\left\{2c_1\frac{F_1'(ka)}{(ka)^3} - \frac{c_2}{(ka)^2}\left(2\frac{F_2'(ka)}{ka} + 1\right)\right\}\cos\omega t$$

$$+ \frac{\pi a^4}{\mu}\left\{2c_2\frac{F_1'(ka)}{(ka)^3} + \frac{c_1}{(ka)^2}\left(2\frac{F_2'(ka)}{ka} + 1\right)\right\}\sin\omega t,$$

and in figure 5.1 we show appropriate coefficients of Q as a function of ka.

Sanyal (1956) has considered pipe flows for which the pressure gradient is either increasing or decreasing exponentially with time. The treatment of the problem closely follows that above, with ω taken to be a purely imaginary quantity.

5.1.3 Pulsed pipe flow

Lance (1956) considered the modifications to Poiseuille flow in a pipe of radius a when the pipe itself is subjected to a series of axial pulses. With v_z represented as in equation (5.5), $\tilde{v}_z$ satisfies (5.6) but with $c \equiv 0$. The boundary conditions for $\tilde{v}_z$ now require

$$\tilde{v}_z(r, t) = 0, \quad t < 0; \quad \tilde{v}_z(a, t) = \sum_{n=0}^{N} U_n\delta(t - nt_0), \quad t > 0.$$

So, for $t > 0$ the pipe wall is subjected to a series of pulses, each represented by a delta function, at intervals t_0 and of strength U_n. If $\bar{\tilde{v}}_z$ denotes the Laplace transform of $\tilde{v}_z$, then $\bar{\tilde{v}}_z$ satisfies

$$r\frac{d^2\bar{\tilde{v}}_z}{dr^2} + \frac{d\bar{\tilde{v}}_z}{dr} - \frac{rs}{\nu}\bar{\tilde{v}}_z = 0, \tag{5.8}$$

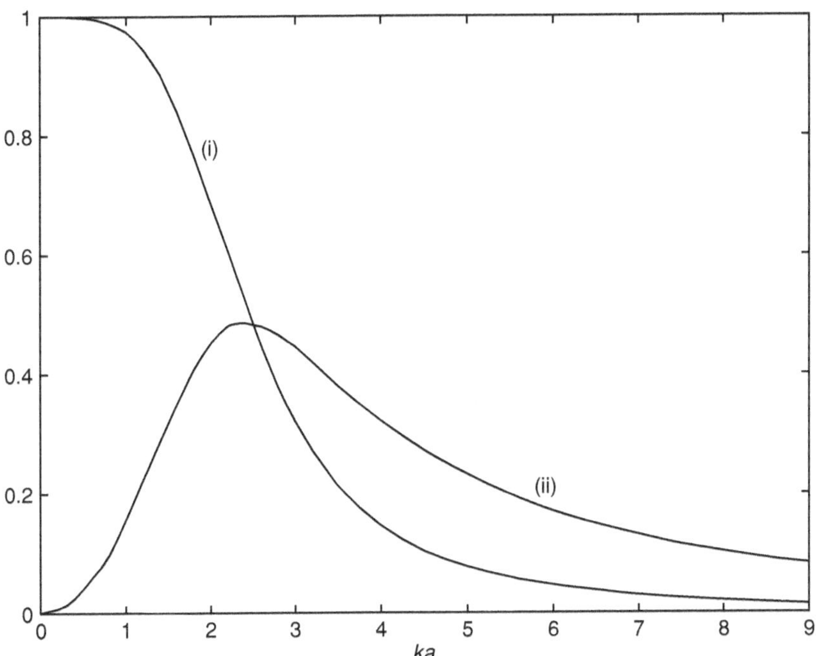

Figure 5.1 Coefficients in the volume flux Q: (i) $16F_1'(ka)/(ka)^3$, (ii) $8\{2F_2'(ka)/ka + 1\}/(ka)^2$.

together with

$$\bar{\bar{v}}_z = \sum_{n=0}^{N} U_n \, e^{-nt_0 s}, \quad \text{at} \quad r = a, \tag{5.9}$$

and a boundedness condition at $r = 0$. The solution of (5.8) with (5.9) for $\bar{\bar{v}}_z$ is

$$\bar{\bar{v}}_z = \frac{J_0\{ir(s/\nu)^{1/2}\}}{J_0\{ia(s/\nu)^{1/2}\}} \sum_{n=0}^{N} U_n \, e^{-nt_0 s}.$$

As a consequence the transform of the volume flux of the unsteady part of the flow field is

$$\bar{Q} = 2\pi \int_0^a r\bar{\bar{v}}_z \, dr = -2\pi i a \left(\frac{\nu}{s}\right)^{1/2} \frac{J_1\{ia(s/\nu)^{1/2}\}}{J_0\{ia(s/\nu)^{1/2}\}} \sum_{n=0}^{N} U_n \, e^{-nt_0 s},$$

and from this, the instantaneous volume flux across any section is given by

$$Q = \frac{P\pi a^4}{8\mu} + \frac{2\pi a^4}{\nu}\frac{\partial}{\partial t}\left\langle \sum_{n=0}^{N} U_n \left[\frac{1}{2} - 2\sum_{m=1}^{\infty} \lambda_m^{-2} \right.\right.$$

$$\left.\left. \times \exp\left\{ -\frac{\lambda_m^2 \nu}{a^2}(t - nt_0) \right\} \right] H(t - nt_0) \right\rangle,$$

where λ_m is the mth zero of J_0. If Q_0 is the volume flux for a single pulse then, when this has been established,

$$Q_0 = \frac{P\pi a^4}{8\mu} + 4\pi a^2 U_0 \sum_{m=1}^{\infty} \exp\left(-\frac{\lambda_m^2 \nu}{a^2}t \right). \tag{5.10}$$

From (5.10) it is seen that for a range of values of t, $0 < t < t_1$, $Q_0 < 0$ provided that

$$\frac{32 U_0 \mu}{P a^2} \sum_{m=1}^{\infty} \exp\left(-\frac{\lambda_m^2 \nu}{a^2}t \right) < -1,$$

that is, provided that the pulse strength is sufficiently large and negative. This consideration motivated, in part, the work of Lance at a time when it was thought that the firing of guns in certain combat aircraft might be the cause of fuel starvation.

5.1.4 The effects of suction or injection on periodic flow

In section 3.4.3 we considered the flow within a porous tube, in particular as studied by Terrill and Thomas (1969). Skalak and Wang (1977) have addressed this same problem, that is with uniform suction, or injection, with velocity V, at the boundary of the pipe, $r = a$, but now modified by the introduction of an oscillatory pressure gradient along it. Thus, for the pressure, we write

$$\frac{p - p_0}{\rho} = \frac{V^2}{R}\left(2f' - \frac{R}{2}\frac{f^2}{\eta} \right) + A\left(\frac{z}{a}\right)^2 + B\left(\frac{z}{a}\right) e^{i\omega t} \tag{5.11}$$

where, when compared with the steady case, we have set the uniform velocity U_0 to zero. With the velocity components

$$v_r = V\eta^{-1/2} f(\eta), \quad v_z = -2V\frac{z}{a}f'(\eta) + \frac{Ba}{4\nu}h(\eta)\, e^{i\omega t}, \tag{5.12}$$

where, as before, $\eta = (r/a)^2$. From (5.11) and (5.12) equation (1.21) gives, for $f(\eta)$, equation (3.35) with boundary conditions (3.36), and for $h(\eta)$,

$$\eta h'' + h' + \tfrac{1}{2}R(f'h - fh') - iM^2 h = 1; \quad R = \frac{Va}{\nu}, \quad M^2 = \frac{\omega a^2}{4\nu}, \tag{5.13}$$

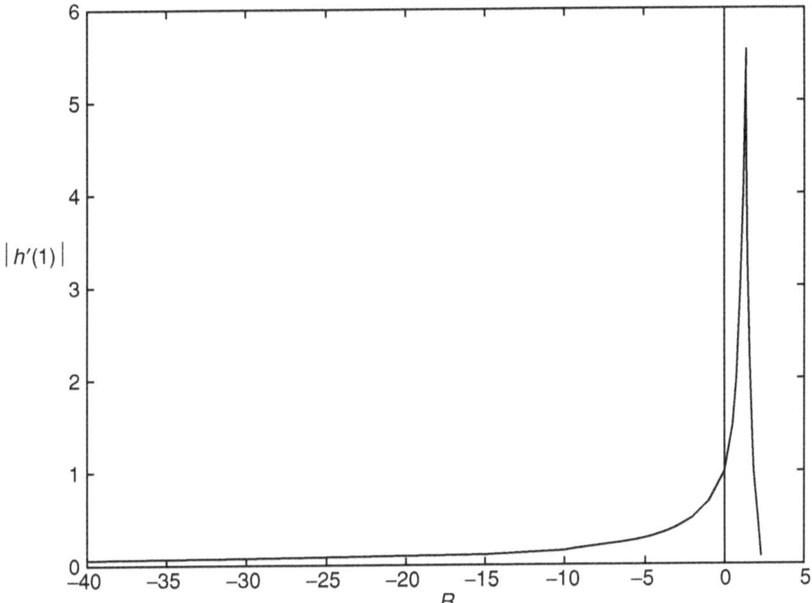

Figure 5.2 The variation of $|h'(1)|$ with R ($M^2 = 0.2$).

together with

$$\lim_{\eta \to 0}(\eta^{1/2}h') = h(1) = 0.$$

The solutions of (3.35), (3.36) for the steady case have been considered in chapter 3, and the complex nature of the solutions is illustrated in figure 3.9. For the unsteady case an additional parameter, M, is involved. Skalak and Wang have made a comprehensive survey of the solutions of (5.13) in (M, R)-parameter space. One of the more interesting features is for small M, and for the solution set I, a resonance type of phenomenon close to $R = 0$. This is illustrated in figure 5.2 where $|h'(1)|$ is plotted as a function of R for $M^2 = 0.2$.

Unsteady annular pipe flow with suction and injection has been considered by Wang (1971). With $v_r = V$ at $r = a$ and $v_r = Va/b$ at $r = b$, $b < a$, the radial component of velocity is simply $v_r = Va/r$, and with the pressure given by

$$\frac{p - p_0}{\rho} = \frac{\nu V a}{2r^2} - \frac{Pz}{\rho} + \frac{\bar{P}z}{\rho}\,\mathrm{e}^{\mathrm{i}\omega t},$$

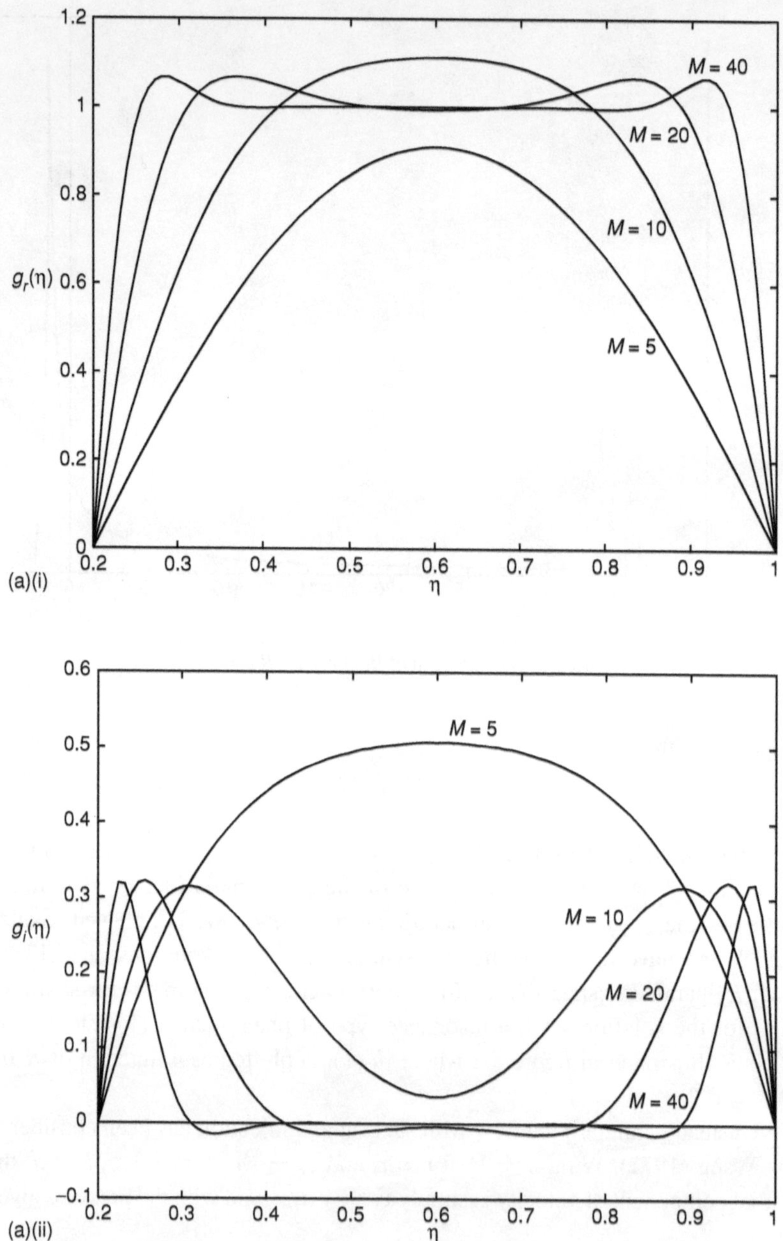

Figure 5.3 (a) Velocity profiles $g = g_r + \mathrm{i}g_i$ for various values of M with $R = 1$. (i) $g_r(\eta)$, (ii) $g_i(\eta)$. (b) As (a) with $R = -1$.

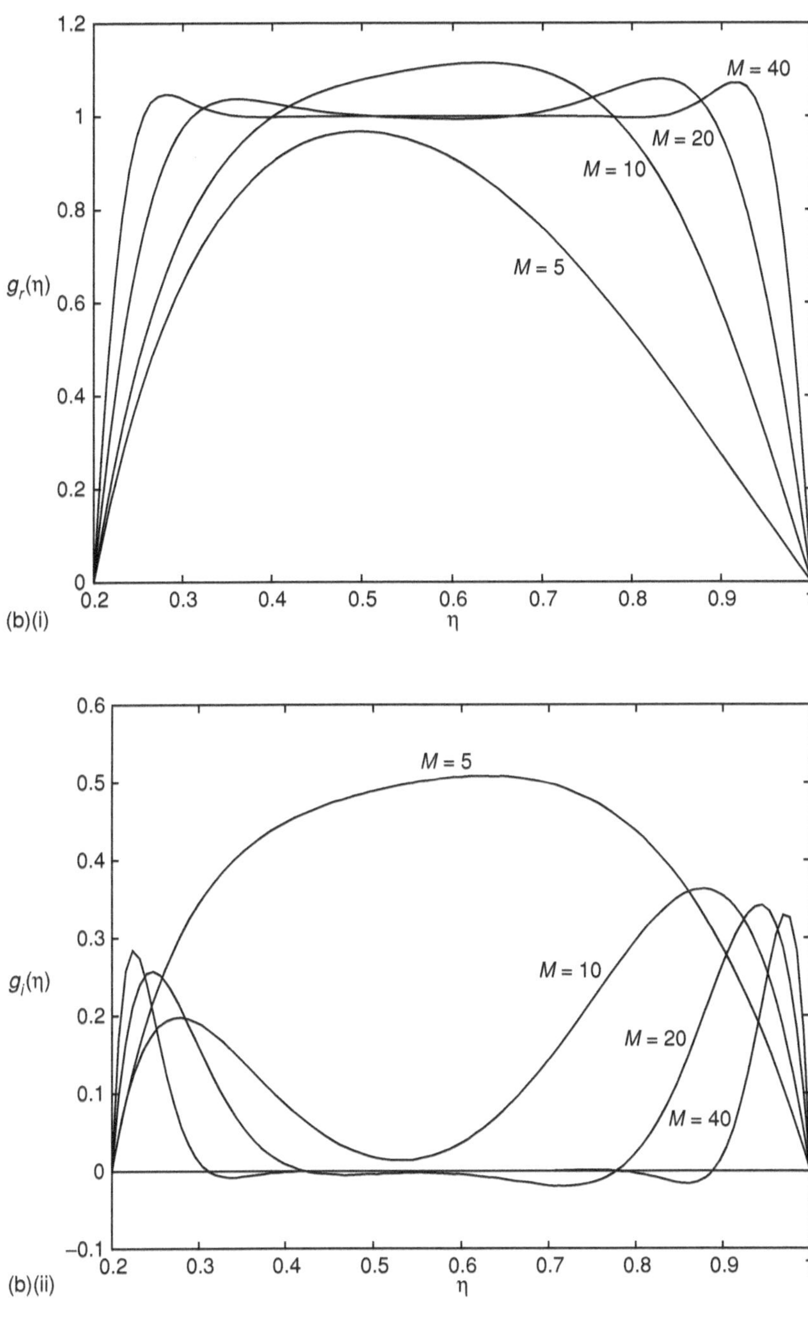

Figure 5.3 (*cont.*)

the axial component of velocity has the form

$$v_z = \{f(\eta) + g(\eta)\, \mathrm{e}^{\mathrm{i}\omega t}\}, \quad \eta = r/a. \tag{5.14}$$

The time-independent part of (5.14) has been discussed by Berman (1958a), and from (1.21) the equation for $g(\eta)$ is

$$g'' + \frac{1}{\eta}(1 - R)g' - \mathrm{i}M^2 g = \frac{\bar{P}a^2}{\mu V}, \quad \text{where now} \quad M^2 = \frac{\omega a^2}{\nu}. \tag{5.15}$$

An exact solution of equation (5.15), with $v_z = 0$ at $\eta = b/a$, 1 which we do not reproduce, involving Bessel functions of imaginary argument, has been obtained by Wang (1971). However, Wang notes that for the special cases $R = 1 - 2n$, where n is an integer, the general solution of (5.15) is

$$g(\eta) = \frac{\mathrm{i}\bar{P}}{\rho \omega V} + C_1 \left(\frac{1}{\eta}\frac{\mathrm{d}}{\mathrm{d}\eta}\right)^n \exp\left\{-\frac{(1+\mathrm{i})}{\sqrt{2}}M\eta\right\}$$

$$+ C_2 \left(\frac{1}{\eta}\frac{\mathrm{d}}{\mathrm{d}\eta}\right)^n \exp\left\{\frac{(1+\mathrm{i})}{\sqrt{2}}M\eta\right\}.$$

In figure 5.3 we show representative velocity profiles for the cases $n = 0$, 1 and various values of M, with $b/a = 0.2$.

5.1.5 Pipes with varying radius

Uchida and Aoki (1977) and Skalak and Wang (1979) have considered the problem of a contracting or expanding pipe. The motivation for the investigation, in part anyway, is physiological with application to an understanding of flows associated with vascoconstriction, for example forced contractions and expansions of a valved vein or a thin bronchial tube. Both sets of authors consider contraction/expansion rates of the form $a(t) = a_0(1 - \alpha t)^{1/2}$ where α may be positive (contraction) or negative (expansion). The similarity reductions differ slightly; following Skalak and Wang we write

$$\psi = \frac{\alpha a_0^2 z}{2} f(\eta) \quad \text{where} \quad \eta = \frac{r^2}{a_0^2(1 - \alpha t)},$$

so that

$$v_r = -\frac{\alpha a_0 f(\eta)}{2(1 - \alpha t)^{1/2}\eta^{1/2}} \quad \text{and} \quad v_z = \frac{\alpha z f'(\eta)}{(1 - \alpha t)}.$$

The pressure p is given by

$$\frac{p_0 - p}{\rho} = \frac{\alpha^2 a_0^2}{4(1 - \alpha t)}\left(\frac{\operatorname{sgn}\alpha}{M^2}f' + \frac{f^2}{2\eta} - f\right) - \frac{2\nu\alpha z^2 A}{a_0^2(1 - \alpha t)^2},$$

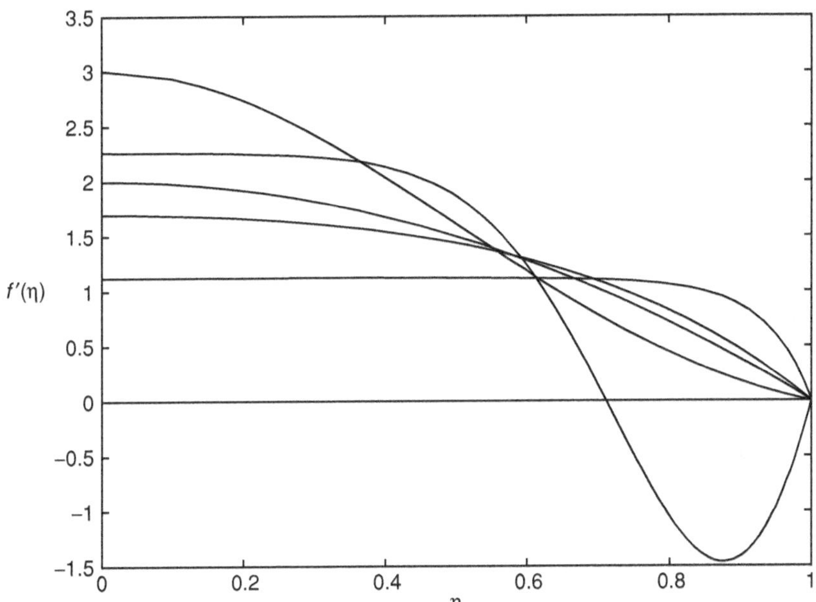

Figure 5.4 Velocity profiles within a contracting ($\alpha > 0$) and expanding ($\alpha < 0$) pipe for various values of M^2. At $\eta = 0$, from the lowest, we have $M^2 = 30.416$ ($\alpha > 0$), $M^2 = 0.86085$ ($\alpha > 0$), $M^2 = 0$, $M^2 = 6.22206$ ($\alpha < 0$), $M^2 = 0.73305$ ($\alpha < 0$).

where A is a constant and p_0 may depend upon t, and equation (1.21) yields, as the equation for $f(\eta)$,

$$\eta f''' + f'' - \operatorname{sgn}\alpha M^2(\eta f'' + f' + f'^2 - f f'') = A$$

where $M^2 = a_0^2 |\alpha|/4\nu$, together with boundary conditions

$$\lim_{\eta \to 0}(\eta^{-1/2} f) = 0, \quad \lim_{\eta \to 0}(\eta^{1/2} f'') = 0$$

which express symmetry of the solution about the pipe centre-line $r = 0$, and

$$f'(1) = 0, \quad f(1) = 1$$

which, respectively, represent no slip at the pipe wall and no flow through it. The constant A is determined as part of the solution procedure as $A = -\operatorname{sgn}\alpha M^2\{f'(0) + f'(0)^2\}$.

In figure 5.4 velocity profiles, for various values of M^2 for both expanding and contracting pipes, are presented.

Typical of the 'squeezing' flows ($\alpha > 0$) is the development of boundary layers at the pipe boundary as M increases. For $\alpha < 0$ flow reversal is encountered as M increases, with outflow occurring close to the pipe wall.

In the above, the radial boundary motion is unidirectional. Blyth, Hall and Papageorgiu (2003) have addressed the problem of flow within the pipe when its radius varies in a time-oscillatory manner. Suppose that we now write

$$\psi = vzf(\eta, t) \quad \text{where} \quad \eta = \frac{r}{a(t)} \tag{5.16}$$

then, since the vorticity has the single component $-(\partial v_z/\partial r)\hat{\boldsymbol{\theta}}$, equation (1.7) gives, for f,

$$\frac{a^2}{v}\left(f_{\eta\eta t} - \frac{f_{\eta t}}{\eta}\right) = \eta\left(\frac{f_\eta}{\eta}\right)_{\eta\eta\eta} + \left(\frac{3a}{v}\frac{da}{dt} - \frac{1}{\eta^2}\right)f_\eta$$

$$+ \left(1 + f + \frac{a}{v}\frac{da}{dt}\eta^2\right)\left(\frac{f_\eta}{\eta}\right)_{\eta\eta} + \left(\frac{f_\eta}{\eta} - f_{\eta\eta}\right)\frac{f_\eta}{\eta},$$

$$\tag{5.17}$$

with boundary conditions

$$\lim_{\eta\to 0}(\eta^{-1}f) = \lim_{\eta\to 0}(\eta^{-1}f_\eta)_\eta = 0; \quad f(1) = -\frac{a}{v}\frac{da}{dt}, \quad f_\eta(1) = 0. \tag{5.18}$$

It is clear from (5.17) and (5.18) that complete self-similarity is obtained only when $a\,da/dt = \text{constant}$ and $\partial f/\partial t = 0$, which corresponds to the case considered by Uchida and Aoki, and Skalak and Wang, treated above. The formulation (5.16) has, however, resulted in a reduction by one of the independent variables. Blyth *et al.* treat, in detail, the case $a(t) = a_0(1 + \Delta\cos\omega t), 0 < \Delta < 1$. A practical reason for this choice is to understand further the physiological application of transmyocardial laser revascularisation that had earlier been partly modelled and analysed by Waters (2001). With the parameter $M^2 = \omega a_0^2/v$ numerical solutions of (5.17) and (5.18) have been obtained, revealing complex dynamics. For small M the flow and forcing are synchronous, and as M increases a Hopf bifurcation occurs. As M increases further the dynamics depend upon Δ. For small Δ the Hopf bifurcation leads to quasi-periodic solutions. At intermediate values of Δ the solutions tend to a chaotic attractor at large t. For larger values of Δ the solution remains periodic as M increases with intervals of integer multiples of the driving period emerging.

5.1.6 Impulsive cylinder flows

The flow induced by a solid cylinder set into a 'sliding' motion parallel to its generators has been considered by several authors. Suppose that the cylinder surface is porous, and at it a velocity of suction $v_r|_{r=a} = -V$ is applied, and let the cylinder be set into motion with speed $w_0(t)$. Then if in the resulting motion $v_z = v_z(r, t)$ we have $v_r = -Va/r$ and, from equation (1.21), the equation for v_z is

$$\frac{\partial v_z}{\partial t} = v \left\{ \frac{\partial^2 v_z}{\partial r^2} + \frac{(1+R)}{r} \frac{\partial v_z}{\partial r} \right\} \quad \text{where} \quad R = \frac{Va}{v}, \tag{5.19}$$

with

$$v_z = 0, t = 0, r > a; \quad v_z = w_0(t), t > 0, r = a; \quad v_z \to 0 \text{ as } r \to \infty.$$

Laplace transform techniques yield as the solution of equation (5.19), with its attendant boundary conditions,

$$v_z(r, t) = \frac{1}{2\pi i} \left(\frac{a}{r} \right)^{R/2} \int_{c-i\infty}^{c+i\infty} e^{st} \frac{K_{R/2}\left(r\sqrt{s/v} \right)}{K_{R/2}\left(a\sqrt{s/v} \right)} \bar{w}_0(s) \, ds, \tag{5.20}$$

where $K_{R/2}$ is a modified Bessel function of the second kind and

$$\bar{w}_0(s) = \int_0^\infty e^{-st} w_0(t) \, dt.$$

The formal solution (5.20) was given by Irmay and Zuzovsky (1970) for the case $R = 0$. For constant sliding velocity, with $w(t) = W$ for $t > 0$, so that $\bar{w}_0 = s^{-1}$, Hasimoto (1956) derives from (5.20) the solution

$$v_z(r, t) = W \left\{ \left(\frac{a}{r} \right)^R + \frac{2}{\pi} \left(\frac{a}{r} \right)^{R/2} \right.$$
$$\left. \times \int_0^\infty \frac{e^{-v\sigma^2 t}}{\sigma} \left(\frac{J_{R/2}(\sigma r)Y_{R/2}(\sigma a) - J_{R/2}(\sigma a)Y_{R/2}(\sigma r)}{J_{R/2}^2(\sigma a) + Y_{R/2}^2(\sigma a)} \right) d\sigma \right\}, \tag{5.21}$$

a solution also given for the case $R = 0$ by Batchelor (1954) by analogy with the corresponding heat-conduction problem of a circular cylinder whose temperature is suddenly raised to a uniform value (see Carslaw and Jaeger (1947)). Batchelor shows, in particular, that for the shear stress at the cylinder

$$\mu \frac{\partial v_z}{\partial r} \bigg|_{r=a} \sim - \left(\frac{\rho \mu}{\pi t} \right)^{1/2} W \quad \text{as} \quad t \to 0.$$

This result coincides with that for the classical impulsive flat plate problem considered in section 4.2, as may be expected when $t \ll a^2/v$.

Hasimoto has observed that considerable simplifications of (5.21) may be achieved for values of $R = -1 + 2n$, where n is an integer. For example with $R = -1$, which corresponds to transpiration across the cylinder surface:

$$v_z = W \operatorname{erfc} \left\{ \frac{r - a}{2(\nu t)^{1/2}} \right\} \quad \text{so that} \quad v_z \to W \text{ as } t \to \infty \text{ for any fixed } r.$$

For the suction case $R = 1$:

$$v_z = \frac{aW}{r} \operatorname{erfc} \left\{ \frac{r - a}{2(\nu t)^{1/2}} \right\} \quad \text{so that} \quad v_z \sim \frac{aW}{r} \text{ as } t \to \infty \text{ for any fixed } r.$$

In a not unrelated problem Lagerstrom and Cole (1955) have considered problems in which the cylinder expands radially with radius $a(t) = At^n$, and again slides parallel to its generators with speed W. In particular they find an exact solution for $n = \frac{1}{2}$. In that case, with $a(t) = a_0(Wt/a_0)^{1/2}$ where a_0 is a constant, the continuity equation $\partial(rv_r)/\partial r = 0$ with $v_r = da/dt$ at $r = a$ gives

$$v_r = \frac{a}{r} \frac{da}{dt} = \frac{a_0 W}{2r}. \tag{5.22}$$

With $v_z = v_z(r, t)$ the momentum equation (1.21) is then

$$\frac{\partial v_z}{\partial t} = \nu \left\{ \frac{\partial^2 v_z}{\partial r^2} + \frac{1}{r} \left(1 - \frac{R}{2} \right) \frac{\partial v_z}{\partial r} \right\}, \quad R = \frac{W a_0}{\nu}, \tag{5.23}$$

which may be compared with equation (5.19). The boundary conditions now require

$$v_z = 0, t = 0, r > 0; v_z = W, t > 0, r = a_0 \left(\frac{Wt}{a_0} \right)^{1/2}; v_z \to 0 \text{ as } r \to \infty.$$

A similarity solution is available by writing $v_z(r, t) = Wf(\eta)$ where $\eta = r/2(\nu t)^{1/2}$ so that f satisfies

$$\eta f'' + \left(1 - \tfrac{1}{2}R + 2\eta \right) f' = 0, \quad f(\sqrt{R}/2) = 1, \quad f(\infty) = 0;$$

with solution

$$f(\eta) = \int_\eta^\infty e^{-\sigma^2} \sigma^{R/2-1} d\sigma \bigg/ \int_{\sqrt{R}/2}^\infty e^{-\sigma^2} \sigma^{R/2-1} d\sigma. \tag{5.24}$$

As in the previous example the expression (5.24) simplifies for particular values of the Reynolds number R, namely $R = 2n$ for integer n. For example with $R = 2, 4$ we have, respectively,

$$v_z = W \frac{\operatorname{erfc} \eta}{\operatorname{erfc}(1/\sqrt{2})}, \quad v_z = W e^{(1-\eta^2)}.$$

In this original formulation the expanding cylinder has $r = 0$ at $t = 0$. For a slight generalisation write $a(t) = a_0\{(Wt + a_0)/a_0\}^{1/2}$, $t > 0$ so that at $t = 0$, $a(0) = a_0$. With $\eta = r/2\{v(t + a_0/W)\}^{1/2}$ the solution (5.24) is unchanged.

In the above examples the cylinder in motion has been of circular cross-section. Tranter (1951) has considered the case of an impermeable cylinder of elliptic cross-section set into uniform motion at the initial instant parallel to its generators. A formal solution is obtained by introducing elliptic co-ordinates, which coincides with Batchelor's solution in the limit as the ellipse axis ratio approaches unity.

In addition to impulsive sliding flows, an exact solution is available for the case in which a circular cylinder of radius a is set into motion with uniform angular velocity V/a. With $v_\theta = v_\theta(r, t)$ equation (1.20) gives

$$\frac{\partial v_\theta}{\partial t} = v\left(\frac{\partial^2 v_\theta}{\partial r^2} + \frac{1}{r}\frac{\partial v_\theta}{\partial r} - \frac{v_\theta}{r^2}\right), \tag{5.25}$$

together with

$$v_\theta = 0, r > a, t = 0; \quad v_\theta = V, r = a, t > 0; \quad v_\theta \to 0 \text{ as } r \to \infty, t \geq 0.$$

The solution of the Laplace transform of (5.25) is again readily obtained, and was given by Lagerstrom (1996) as

$$\bar{v}_\theta = \frac{V}{2\pi i}\int_{c-i\infty}^{c+i\infty} \frac{e^{st}}{s}\frac{K_1\left(r\sqrt{s/v}\right)}{K_1\left(a\sqrt{s/v}\right)}\,ds. \tag{5.26}$$

The inverse of (5.26) is

$$v_\theta = V\left\{\frac{a}{r} + \frac{2}{\pi}\int_0^\infty \frac{e^{-v\sigma^2 t}}{\sigma}\left(\frac{J_1(\sigma r)Y_1(\sigma a) - J_1(\sigma a)Y_1(\sigma r)}{J_1^2(\sigma a) + Y_1^2(\sigma a)}\right)d\sigma\right\}. \tag{5.27}$$

The similarity between these Rayleigh-type sliding and rotating flows is evidenced by a comparison of equations (5.21) and (5.27).

Oscillatory flows, the analogue of Stokes' problem for the oscillating flat plate discussed in chapter 4, also admit exact solutions for the circular cylinder of radius a. Consider first the situation in which an impermeable cylinder performs an oscillatory motion parallel to its generators such that at $r = a$, $v_z = W e^{i\omega t}$. Equation (5.19), with $R \equiv 0$, is the equation satisfied by $v_z(r, t)$ with solution

$$v_z = W\frac{K_0\left\{r(i\omega/v)^{1/2}\right\}}{K_0\left\{a(i\omega/v)^{1/2}\right\}}\,e^{i\omega t}. \tag{5.28}$$

Far from the cylinder equation (5.28) gives

$$v_z \sim C r^{-1/2} \exp\{-(\omega/2\nu)^{1/2}r\} \exp[\mathrm{i}\{\omega t - (\omega/2\nu)^{1/2}r\}] \quad \text{as} \quad r \to \infty,$$
(5.29)

which represents a decaying viscous wave as in the two-dimensional Stokes solution but with the decay rate now enhanced by a factor $r^{-1/2}$ reflecting the change of geometry. Hasimoto (1956) has also addressed this problem when suction is applied at the cylinder surface.

When the cylinder performs torsional oscillations with angular velocity $(V/a)\,\mathrm{e}^{\mathrm{i}\omega t}$ equation (5.25) is satisfied by v_θ with solution, for the cylinder assumed impermeable,

$$v_\theta = V \frac{K_1\left\{r(\mathrm{i}\omega/\nu)^{1/2}\right\}}{K_1\left\{a(\mathrm{i}\omega/\nu)^{1/2}\right\}}\mathrm{e}^{\mathrm{i}\omega t},$$

which also exhibits the behaviour (5.29) as $r \to \infty$.

5.2 Beltrami flows and their generalisation

As we have noted in earlier chapters, Beltrami flows are those for which $\mathbf{v} \wedge \boldsymbol{\omega} \equiv \mathbf{0}$ and it was shown, in particular, in section 4.6 that there can be no such unsteady flows that are axisymmetric. For their generalisation, with $\nabla \wedge (\mathbf{v} \wedge \boldsymbol{\omega}) \equiv \mathbf{0}$, $\boldsymbol{\omega}$ satisfies equation (4.39), the solution of which for axisymmetric flows does not appear to have attracted close attention. However, Irmay and Zuzovsky (1970) have presented one such solution for which the stream function is given as

$$\psi = -(c_1 + c_2 z)\,\mathrm{e}^{-r^2/4\nu t},$$

where c_1, c_2 are constants, so that

$$v_r = \frac{c_2}{r}\,\mathrm{e}^{-r^2/4\nu t}, \quad v_z = \frac{c_1 + c_2 z}{2\nu t}\,\mathrm{e}^{-r^2/4\nu t} \quad \text{and} \quad \boldsymbol{\omega} = \frac{(c_1 + c_2 z)r}{4\nu^2 t^2}\,\mathrm{e}^{-r^2/4\nu t}\,\hat{\boldsymbol{\theta}}.$$
(5.30)

From (5.30) it is seen that at $t = 0$ the fluid is everywhere at rest. For $t > 0$ a radial flow develops from the axis $r = 0$, together with an axial flow induced by an appropriate pressure gradient. As $t \to \infty$, $v_z \to 0$ for $r > 0$, and the flow becomes the irrotational flow due to a line source on the axis $r = 0$.

5.3 Stagnation-point flows

As for the unsteady two-dimensional problems considered in chapter 4 there are unsteady axisymmetric, and indeed three-dimensional, exact solutions of the Navier–Stokes equations constructed by superposing a time-dependent motion, typically a fluctuation, upon a known steady solution. We begin by considering such a case.

5.3.1 The Homann flow against an oscillating plate

Weidman and Mahalingham (1997) consider the axisymmetric stagnation-point flow against a porous plane boundary at which the transpiration velocity is a constant, equal to $-W_0$, and which performs periodic oscillations in its own plane, in the x-direction, with frequency ω. With rectangular co-ordinates (x, y, z) the velocity components may be written as

$$u = kxf'(\eta) + U_0 g(\eta)\,\mathrm{e}^{i\omega t}, \quad v = kyf'(\eta), \quad w = -2(kv)^{1/2}f(\eta) - W_0,$$

$$\eta = \left(\frac{k}{v}\right)^{1/2} z, \tag{5.31}$$

and the pressure as

$$\frac{p - p_0}{\rho} = -\frac{1}{2}k^2(x^2 + y^2) - 2kv\{f^2(\eta) + f'(\eta)\} - 2(kv)^{1/2}W_0. \tag{5.32}$$

In equation (5.31) k is again the constant strain rate with U_0/ω the amplitude of the plate oscillations. As in the two-dimensional case the pressure is unaffected by the oscillatory component of velocity. This is evident when comparing (5.32) with equation (3.20) where the difference is accounted for by the suction velocity imposed at the boundary. The velocity field (5.31) satisfies the continuity equation (1.12) and equations (1.10) and (1.11) are satisfied identically provided that f, g satisfy the ordinary differential equations

$$f''' + 2ff'' + Sf'' - f'^2 + 1 = 0, \tag{5.33}$$

$$g'' + 2fg' + Sg' - f'g + i\Omega g = 0, \tag{5.34}$$

where $S = W_0/(kv)^{1/2}$ and $\Omega = \omega/k$. The boundary conditions require

$$f(0) = f'(0) = 0, \quad f'(\infty) = 1; \quad g(0) = 1, \quad g(\infty) = 0. \tag{5.35}$$

There are special cases: (a) $U_0 = S = 0$, which corresponds to the classical steady, axisymmetric stagnation-point flow of Homann (1936), (b) $\Omega = S = 0$, which represents the case of a flat plate moving transversely at uniform speed U_0 beneath the axisymmetric stagnation-point flow as considered by Wang (1973) and (c) $S = 0$, representing the stagnation-point flow towards an impermeable plate performing transverse oscillations.

Weidman and Mahalingam have integrated the system (5.33) to (5.35) numerically for a wide range of values of the parameters S and Ω, and have in

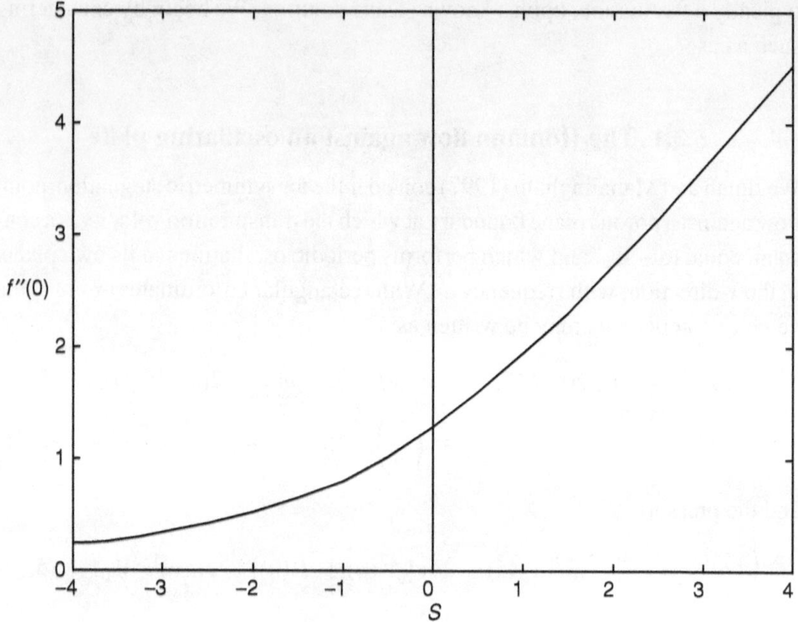

Figure 5.5 The variation of the radial shear-stress parameter $f''(0)$ with S.

addition carried out asymptotic analyses for large $|S|$, Ω. The shear stress on the oscillating plate is given by

$$\boldsymbol{\tau} = \rho k^{3/2} \nu^{1/2} r f''(0)\hat{\mathbf{r}} + \rho U_0 (k\nu)^{1/2} |g'(0)| \, e^{i(\omega t + \phi)}\mathbf{i}$$

where $r = (x^2 + y^2)^{1/2}$ and $\phi = \arg\{g'(0)\}$. Thus, the shear stress has a steady radial component and an unsteady component along the direction of the plate oscillation. In figure 5.5 the dependence of $f''(0)$ upon the transpiration parameter S is shown. As the suction increases so the shear stress increases. For negative values of S, which corresponds to blowing from the plate, the time-independent part of the solution is essentially that of opposing stagnation-point flow with a free stagnation point. The stand–off distance of this point increases with the blowing rate as shown in figure 5.6. For each value of Ω, $|g'(0)|$ also increases as S increases as can be seen in figure 5.7, where $|g'(0)|$ is shown as a function of Ω for discrete values of S. For each value of S, $|g'(0)|$ is seen to increase monotonically with Ω, consistent with the two-dimensional flow discussed in chapter 4. The phase angle ϕ also increases with Ω, for each value of S, as shown in figure 5.8; Weidman and Mahalingam demonstrate that, for all S, $\phi \to \pi/4$ as $\Omega \to \infty$.

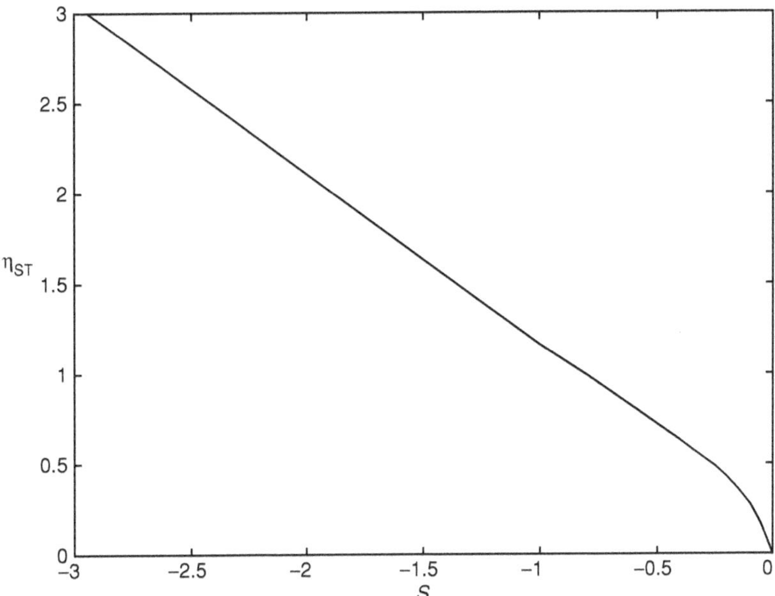

Figure 5.6 The stand-off distance η_{ST} of the free stagnation point as a function of the blowing rate S.

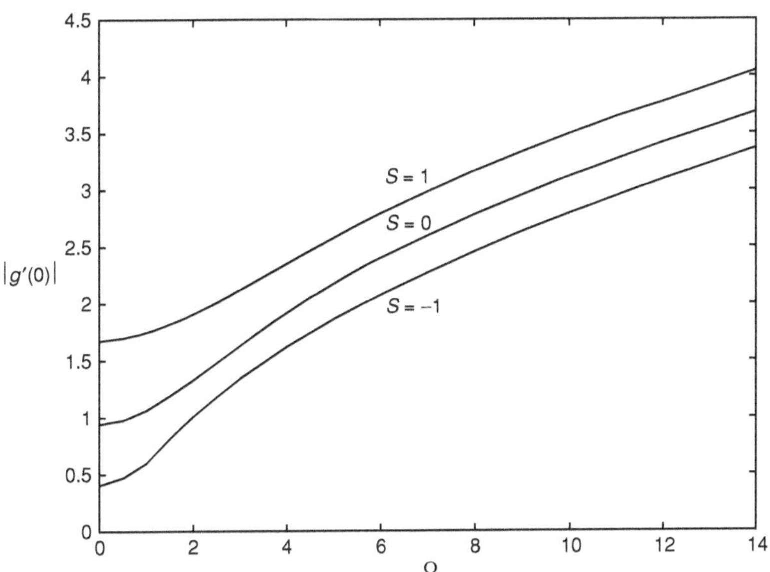

Figure 5.7 The transverse shear-stress parameter $|g'(0)|$ as a function of Ω for various values of S.

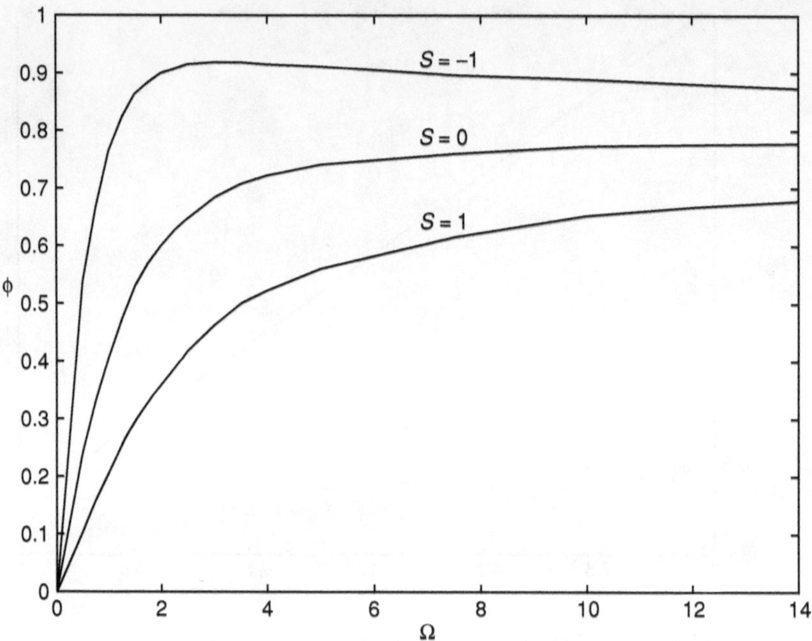

Figure 5.8 The phase angle ϕ as a function of Ω for various values of S.

5.3.2 Oblique stagnation-point flow

In section 4.7.4 we considered the unsteady three-dimensional stagnation-point flow of Cheng *et al.* (1971), where the parameter r $(0 \leq r \leq 1)$ characterises the flow in the sense that for $r = 0$ the flow is two-dimensional, whilst for $r = 1$ an axisymmetric flow is recovered. Wang (1985b) has generalised this solution by introducing a time-dependent cross-flow, in the x-direction, of uniform shear. Although Wang formulates the problem in the general form of Cheng *et al.*, attention is focused upon the case $r = 1$. Following Wang we write

$$u = k\left\{\frac{xf'(\eta)}{1 - kt} + \left(\frac{\nu}{k}\right)^{1/2}\frac{\gamma g(\eta)}{(1 - kt)^{1/2}}\right\}, \quad v = k\frac{yf'(\eta)}{1 - kt},$$

$$w = -2k\left(\frac{\nu}{k}\right)^{1/2}\frac{f(\eta)}{(1 - kt)^{1/2}}, \tag{5.36}$$

where $\eta = (k/\nu)^{1/2} z/(1 - kt)^{1/2}$, and the pressure as

$$\frac{p - p_0}{\rho} = -\frac{k^2(x^2 + y^2)}{(1 - kt)^2} + \frac{\gamma(\nu k^3)^{1/2}}{2(1 - kt)^{3/2}}(3B - 4A)x$$

$$+ \frac{\nu k}{(1 - kt)}\{\eta f(\eta) - 2f^2(\eta) - 2f'(\eta)\}. \tag{5.37}$$

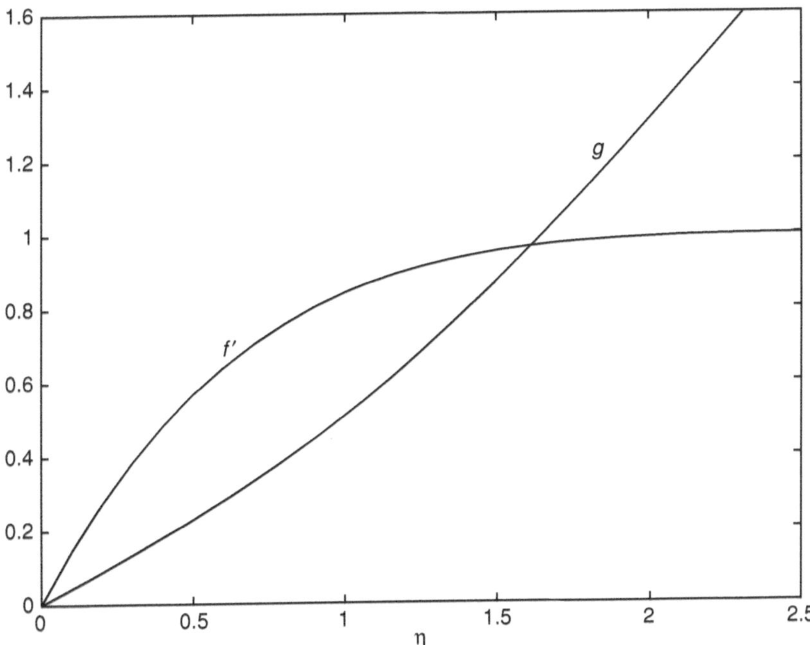

Figure 5.9 The velocity profiles $f'(\eta)$, $g(\eta)$ for the case $B = 4A/3$.

In this expression for the pressure, (5.37), the constant A is determined from the displacement of the basic onset flow, that is $A = \lim_{\eta \to \infty}(\eta - f)$. The constant B is arbitrary, but in turn determines the displacement of the cross-flow, such that $B = \lim_{\eta \to \infty}(\eta - g)$. Wang considers in detail the case for which $B = 4A/3$. With the equation of continuity satisfied identically, equations (1.9) and (1.10) give, using (5.36) and (5.37), as equations for $f(\eta)$ and $g(\eta)$,

$$f''' - \tfrac{1}{2}\eta f'' + 2ff'' - f'^2 - f' + 2 = 0,$$

$$g'' - \tfrac{1}{2}\eta g' + 2fg' - f'g - \tfrac{1}{2}g = 0,$$

with boundary conditions

$$f(0) = f'(0) = g(0) = 0; \qquad f'(\infty) = g'(\infty) = 1.$$

From the solution of these equations the profiles $f'(\eta)$, $g(\eta)$ are shown in figure 5.9.

The components of shear stress at the plate are given by

$$\tau_x = \mu \frac{\partial u}{\partial z}\bigg|_{z=0} = (\rho \mu k^3)^{1/2} \left\{ \frac{x f''(0)}{(1 - kt)^{3/2}} + \left(\frac{\nu}{k}\right)^{1/2} \frac{\gamma g'(0)}{(1 - kt)^{1/2}} \right\},$$

$$\tau_y = \mu \frac{\partial v}{\partial z}\bigg|_{z=0} = (\rho \mu k^3)^{1/2} \frac{y f''(0)}{(1 - kt)^{3/2}}.$$

The vanishing of these two components determines the stagnation point of attachment as, with $f''(0) = 1.5570332$, $g'(0) = 0.4401276$,

$$x_s = -\gamma \left(\frac{\nu}{k}\right)^{1/2} \frac{g'(0)}{f''(0)} (1 - kt) = -0.2827\gamma \left(\frac{\nu}{k}\right)^{1/2} (1 - kt), \quad y_s = 0,$$

which approaches the origin as the singular time $t = k^{-1}$ is approached.

5.3.3 Unsteady stagnation on a circular cylinder

In section 3.4.2 the problem of stagnation flow on a circle of radius a circumscribing a cylinder of the same radius, with generators in the z-direction, was presented. If such a cylinder performs torsional oscillations with angular velocity $(V/a)\, \mathrm{e}^{i\omega_1 t}$ and, simultaneously, longitudinal oscillations with speed $W \mathrm{e}^{i\omega_2 t}$ then, when this is embedded within the steady 'circular' stagnation flow, an exact solution of the three-dimensional Navier–Stokes equations is available with, in cylindrical polar co-ordinates, velocity components given by

$$v_r = -ka\eta^{-1/2} f(\eta), \quad v_\theta = V\eta^{-1/2} g(\eta)\, \mathrm{e}^{i\omega_1 t}, \quad v_z = 2kz f'(\eta) + W h(\eta)\, \mathrm{e}^{i\omega_2 t}, \tag{5.38}$$

where $\eta = (r/a)^2$, and pressure as

$$\frac{p - p_0}{\rho} = -2k^2 z^2 - \frac{1}{2} k^2 a^2 \eta^{-1} f^2(\eta) - 2\nu k f'(\eta)$$

$$+ \frac{1}{2} V^2 \int_1^\eta (|g(s)|/s)^2\, \mathrm{e}^{2i(\omega_1 t + \phi)}\, \mathrm{d}s, \tag{5.39}$$

where $\phi(\eta) = \arg\{g(\eta)\}$. If, in (5.38), we set $V \equiv 0$ then the flow is the analogue of the two-dimensional flow considered by Glauert (1956) and Rott (1956) in which, at a two-dimensional stagnation point, the flat plate oscillates orthogonal to the stagnation line, whilst if W is set to zero the resulting flow is the analogue of Rott's (1956) solution where the plate oscillates along the stagnation line.

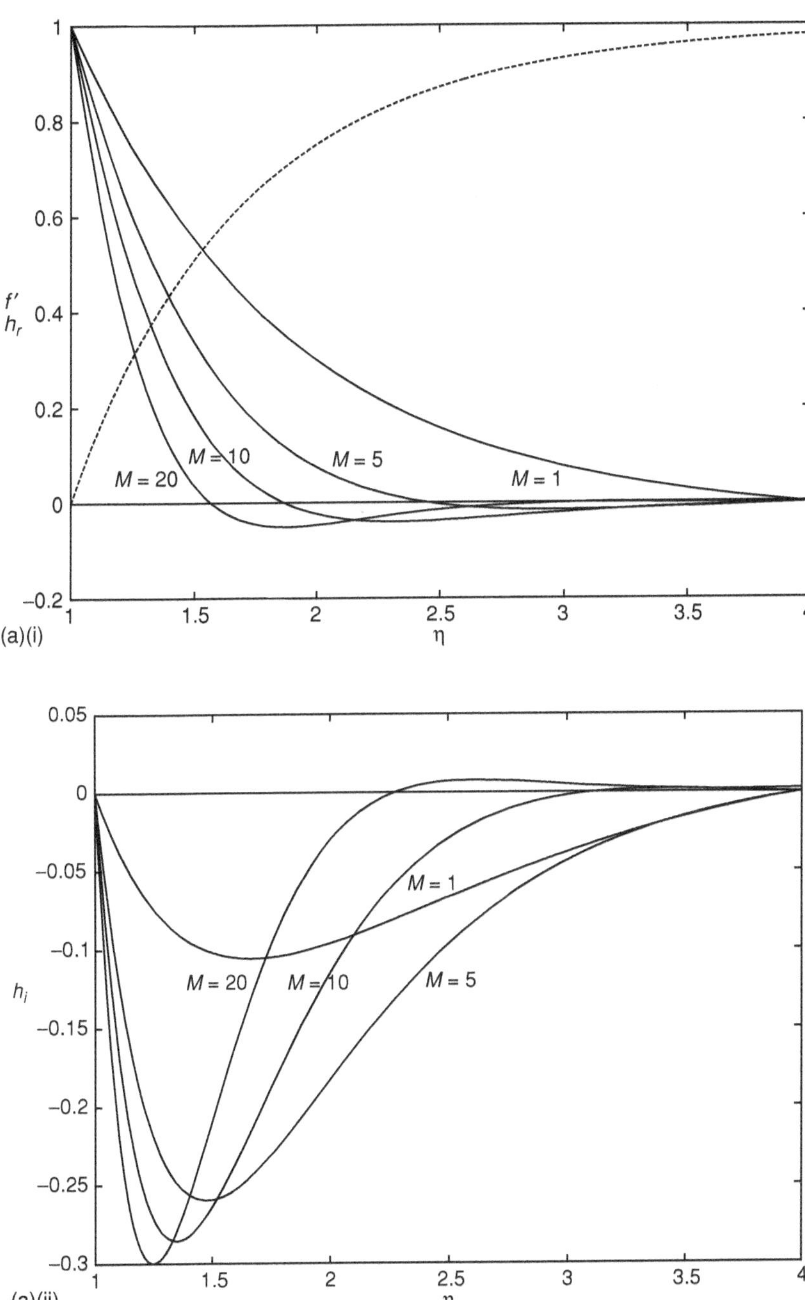

Figure 5.10 (a) Velocity profiles $f'(\eta)$ (broken line) and $h = h_r + \mathrm{i}h_\mathrm{i}$ for various values of M with $R = 1$. (i) $h_r(\eta)$, (ii) $h_i(\eta)$. (b) As (a) with $R = 10$.

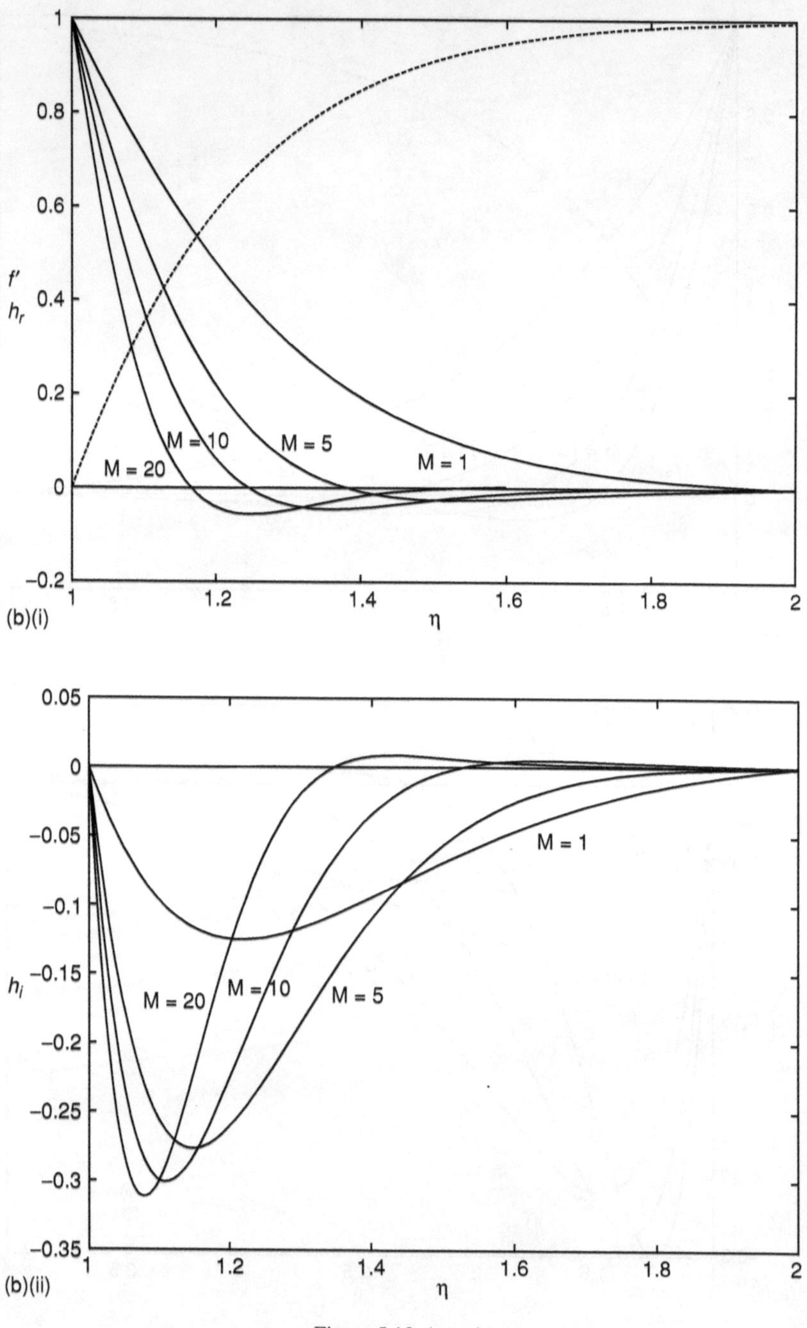

Figure 5.10 (*cont.*)

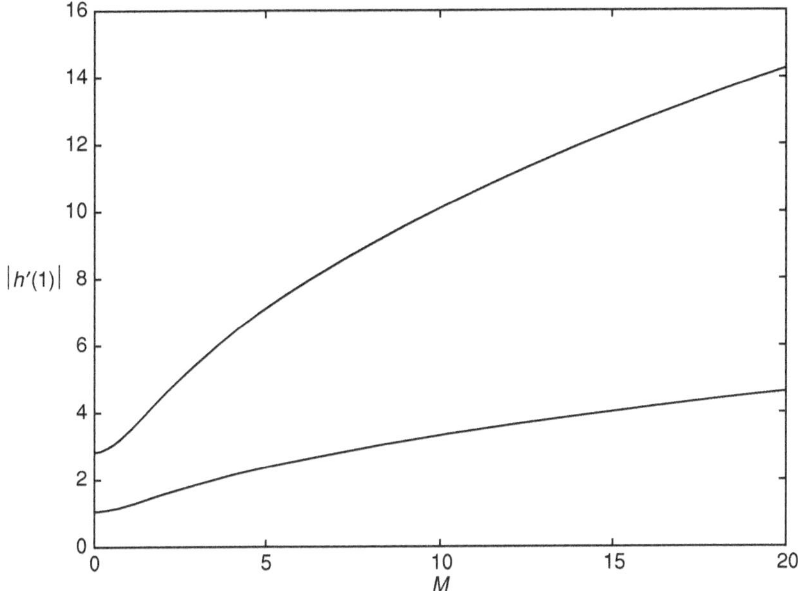

Figure 5.11 Variation of the shear-stress parameter $|h'(1)|$ with M for the two Reynolds numbers $R = 1$ (lower) and $R = 10$ (upper).

With the velocity field and pressure as in equations (5.38) and (5.39), equation (1.22) is satisfied identically whilst (1.19), (1.20) and (1.21) will be satisfied provided that $f(\eta)$, $g(\eta)$ and $h(\eta)$ satisfy the following ordinary differential equations,

$$\eta f''' + f'' + R(ff'' - f'^2 + 1) = 0, \quad f(1) = f'(1) = 0, \quad f'(\infty) = 1;$$

$$\eta g'' + R(fg' - iM_1 g) = 0, \quad g(1) = 1, \quad g(\infty) = 0;$$

$$\eta h'' + h' + R(fh' - f'h - -iM_2 h) = 0, \quad h(1) = 1, \quad h(\infty) = 0.$$

In these equations $R = ka^2/2\nu$ is a Reynolds number and $M_i = \omega_i/2k$ ($i = 1, 2$). The special case $V \equiv 0$ has been considered by Gorla (1979) who determines solutions in the range $10^{-2} \le R \le 10^2$ for various values of M_2. In figure 5.10 we show representative velocity profiles for two values of the Reynolds number, whilst in figure 5.11 the variation of $|h'(1)|$ with M_2 is shown.

5.4 Squeeze flows

The unsteady squeezing of a viscous fluid between two parallel disks provides a model for the loading of mechanical parts as, for example, in a thrust bearing, or for the draining of a fluid layer from between the two planes as in the draining of squeeze films between electrically or thermally conducting surfaces. For axisymmetric flow the fluid motion between the two disks, a distance $h(t)$ apart with $h_0 = h(0)$, has the nature of a double axisymmetric stagnation-point flow. In that case, using cylindrical polar co-ordinates and assuming that the origin is located on the lower disk which is assumed to be fixed, if we write $\psi = -Ur^2 F(z, t)$ with U a typical velocity then

$$v_r = Ur\frac{\partial F}{\partial z} \quad \text{and} \quad v_z = -2UF. \tag{5.40}$$

The continuity equation is satisfied identically and equations (1.19) and (1.21) yield for the unknown F, and the pressure p

$$\frac{1}{R}F_{\eta\eta\eta} + 2FF_{\eta\eta} - F_\eta^2 - F_{\eta\tau} = A(\tau), \tag{5.41}$$

where $\eta = z/h_0$, $\tau = Ut/h_0$, $R = Uh_0/v$,

$$\frac{p - p_0}{\rho} = \frac{1}{2}\frac{U^2r^2}{h_0^2}A(\tau) - 2U^2\left(\frac{1}{R}\frac{\partial F}{\partial \eta} + F^2 + \int_\eta^1 \frac{\partial F}{\partial \tau}\,\mathrm{d}\eta\right), \tag{5.42}$$

where A and the time-dependent reference pressure p_0 are unknown *a priori*. The boundary conditions for equation (5.41) require

$$F = F_\eta = 0 \quad \text{on} \quad \eta = 0; \tag{5.43}$$

$$F_\eta = 0, \quad F = -\frac{1}{2}\frac{\mathrm{d}\bar{h}}{\mathrm{d}\tau}, \quad \text{on} \quad \eta = \bar{h} = h/h_0; \tag{5.44}$$

$$F = 0 \quad \text{and} \quad \bar{h} = 1 \quad \text{at} \quad \tau = 0. \tag{5.45}$$

Two cases of special interest are: (i) when the fluid motion is induced by a constant downward force squeezing the fluid out radially and (ii) when the time-dependent gap width is prescribed. We consider each case in turn.

5.4.1 Constant force

The draining of fluid from the gap between the disks has been considered by Weinbaum, Lawrence and Kuang (1985) and Lawrence, Kuang and Weinbaum (1985) when a constant force is applied to the upper disk. Such a force may simply be the weight W of the upper disk, if it is assumed finite with radius a. In that case $h_0 \ll a$ must be assumed, and the solution (5.40), (5.42) will be appropriate everywhere except within a distance $O(h_0)$ from the edge. If the assumption is made that $h_{tt} \ll g$ then the force balance at the upper disk

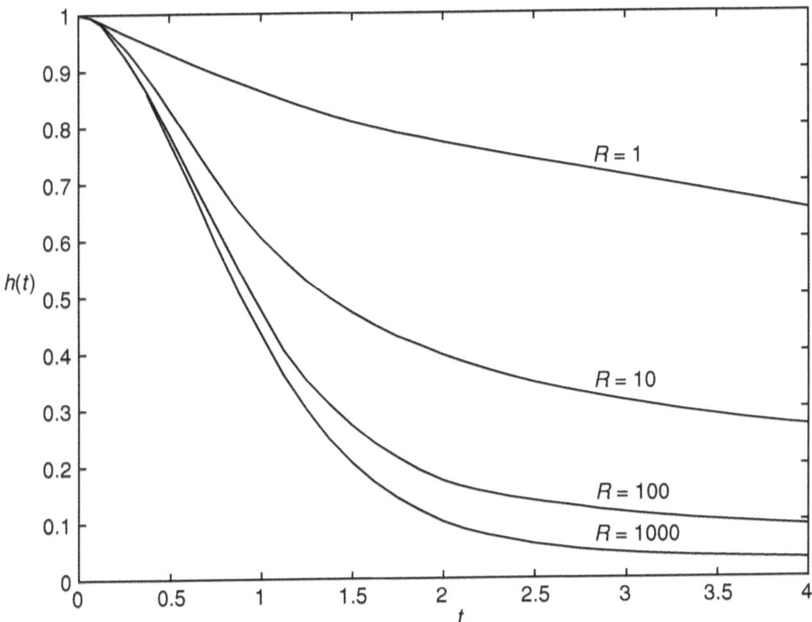

Figure 5.12 The gap width $h(t)$ for various values of the Reynolds number R.

requires $F_p = W$ where F_p is the force due to pressure at $z = h$ which, from (5.42), (5.44) is

$$F_p = \int_0^a 2\pi r p(r, h, t)\, dr = \frac{\pi}{4}\rho U^2 \frac{a^4}{h_0^2} A + \rho \pi a^2 \left\{ p_0 - \frac{1}{2}U^2 \left(\frac{d\bar{h}}{d\tau}\right)^2 \right\},$$

$$(5.46)$$

since $p(r, h, t) = \rho \left\{ U^2 r^2 A/2h_0^2 - U^2 \bar{h}_\tau^2/2 + p_0/\rho \right\}$. If the small drop in pressure across the exit is ignored, an approximation accurate to $O(h_0/a)$, and p is measured from the ambient pressure then $p_0/\rho = U^2 \bar{h}_\tau^2/2 - U^2 a^2 A/2h_0^2$ and equation (5.46) then yields

$$F_p = -\frac{\pi \rho U^2 a^4 A}{4h_0^2} = W. \qquad (5.47)$$

Weinbaum *et al.* have identified a characteristic settling velocity as $2(h_0/a) \times (W/\pi\rho a^2)^{1/2}$ and adopting this as the typical velocity U gives, from (5.47), $A(\tau) \equiv -1$. With this value for A, Lawrence *et al.* have integrated equation (5.41) numerically, subject to the conditions (5.43) to (5.45). Principal amongst the results is the variation of $\bar{h}$ with τ for different values of the Reynolds number R. These are shown in figure 5.12. Predictably the gap closes more rapidly as the Reynolds number increases; for the high Reynolds number limit Lawrence *et al.* show that all the fluid is drained away when $\tau \approx 4$.

5.4.2 Prescribed gap width

Wang (1976) has extended his two-dimensional theory, discussed in section 4.8.2, to this axisymmetric situation. The two disks move relative to one another with the distance between them given as $h(t) = \pm h_0(1 - kt)^{1/2}$, so that initially they are separated by a distance $2h_0$. In this case, with an assumed symmetry of the flow about the mid-plane between the disks, it is convenient to take that plane as the origin for the co-ordinate z. With kh_0 as a characteristic velocity and

$$F(z, t) = \frac{f(\eta)}{4(1 - kt)^{1/2}} \quad \text{where} \quad \eta = \frac{z}{h_0(1 - kt)^{1/2}},$$

equation (5.40) gives

$$v_r = \frac{kr}{4(1 - kt)} f'(\eta), \quad v_z = -\frac{h_0 k}{2(1 - kt)^{1/2}} f(\eta), \tag{5.48}$$

so that, with $A = -A_0/8(1 - kt)^2$, equation (5.41) gives as the ordinary differential equation for f,

$$\frac{2}{R} f''' + ff'' - \frac{1}{2} f'^2 - 2f' - \eta f'' = -A_0, \tag{5.49}$$

and the pressure is determined from (5.42) as

$$\frac{p - p_0}{\rho} = -\frac{k^2 r^2}{16(1 - kt)^2} A_0 - \frac{h_0^2 k^2}{8(1 - kt)} \left(4\frac{f'(\eta)}{R} + f^2(\eta) - 2\eta f(\eta) \right). \tag{5.50}$$

The boundary conditions for equation (5.49) require $f(0) = f''(0) = 0$ from the assumed symmetry about $z = 0$, and $f(1) = 1$, $f'(1) = 0$. That four conditions are required for the third-order equation (5.49) reflects the fact that $A_0 = -2f'''(1)/R$ is not determined *a priori*. In (5.50) the arbitrary time-dependent pressure $p_0(t)$ may be determined if, for example, the disks are assumed to be finite with radius $a \gg h_0$ and the pressure is equated with the ambient pressure, incurring an error $O(h_0/a)$. In particular if $p(a, h, t) = 0$ then

$$p_0(t) = \frac{\rho k^2 a^2}{16(1 - kt)^2} \left\{ A_0 - 2 \left(\frac{h_0}{k} \right)^2 (1 - kt) \right\},$$

and

$$p(r, h, t) = \frac{\rho k^2 (r^2 - a^2) f'''(1)}{16R(1 - kt)^2}. \tag{5.51}$$

Wang has obtained solutions of equation (5.49) for a range of values of R, both positive and negative. Qualitatively the results are similar to the two-dimensional case shown in figures 4.14 and 4.15, and not reproduced here. For $R > 0$, that is $k > 0$, $f'(\eta) > 0$ and fluid is squeezed radially outwards with increasing speed until the disks touch at $t = k^{-1}$. For negative k, and hence negative R, the disks are drawn apart. When $|R|$ is sufficiently small there is a unidirectional radial inflow between the disks with $f''(1) < 0$. However, when $R = R_{c_1} \approx -6.5$, $f''(1) = 0$, and for $R < R_{c_1}$ regions of radial outflow develop in the neigbourhood of each disk with a local maximum that increases with $|R|$ as in the two-dimensional case.

For a finite disk of radius a the pressure force F_p on the upper disk $\eta = 1$ is given, using (5.51), as

$$F_p = \int_0^a 2\pi r p(r, h, t)\, dr = -\frac{\pi \rho k^2 a^4 f'''(1)}{32 R(1 - kt)^2}. \tag{5.52}$$

For $R > 0$ Wang finds that $f'''(1) < 0$, so that there is a net thrust on the two disks resisting their inward motion. For $R < 0$ Wang notes that for $R > R_{c_2} \approx -1.74$, $f'''(1) < 0$ but that for $R < R_{c_2}$, $f'''(1) > 0$. This implies that as the disks move apart there is a suction on them again resisting the motion for $R > R_{c_2}$, but for $R < R_{c_2}$ the force F_p in (5.52) is positive implying an outward thrust on the disks. However, care must be exercised in interpreting these results since for radial inflow, as when $R < 0$, the assumed self-similar form of solution cannot hold everywhere for a finite disk since no conditions at $r = a$ are imposed on the solution, and its range of validity is unknown.

The squeeze flows discussed above are axisymmetric. Wang and Watson (1979) consider squeeze flows in which this axisymmetry of flow properties is abandoned. With the gap width again prescribed as $h(t) = \pm h_0(1 - kt)^{1/2}$ and the similarity variable $\eta = z/h_0(1 - kt)^{1/2}$, the velocity components are written in a rectangular co-ordinate system as

$$u = \frac{kx}{4(1 - kt)} f_1'(\eta), \quad v = \frac{ky}{4(1 - kt)} f_2'(\eta),$$

$$w = -\frac{kh_0}{4(1 - kt)^{1/2}} \{f_1(\eta) + f_2(\eta)\}, \tag{5.53}$$

for which the corresponding pressure distribution may be written as

$$\frac{p}{\rho} = -\frac{k^2 A_0(x^2 + \alpha^2 y^2 - a^2)}{16(1 - kt)^2}$$

$$- \frac{k^2 h_0^2}{8(1 - kt)} \left\{ \frac{2}{R}(f_1' + f_2') + \frac{1}{4}(f_1 + f_2)^2 - \eta(f_1 + f_2) \right\}. \tag{5.54}$$

Table 5.1. *The two values of A_0, A_{01} and A_{02}, obtained when $R = -1$ for various values of α*

α	A_{01}	A_{02}
0.0	1.78647	−5.19816
0.2	1.75267	−7.72627
0.5	1.62476	−15.9346
0.7	1.50184	−27.0332
1.0	1.30226	−53.9867

The pressure $p(x, y, z, t)$ in equation (5.54) has been written in such a manner that for $a \gg h_0$, $p = 0$ at $x^2 + \alpha^2 y^2 = a^2$. An interpretation of this is that the planes which form the gap from which fluid is being squeezed are the ellipses $x^2 + \alpha^2 y^2 = a^2$, with semi-major and semi-minor axes a, b and $\alpha = a/b$. The equations satisfied by $f_i(\eta)$, $i = 1, 2$ are, from (1.9) and (1.10),

$$\frac{2}{R} f_i''' + \frac{1}{2}(f_1 + f_2)f_i'' - \frac{1}{2} f_i^2 - 2f_i - \eta f_i'' = -A_0,$$

for which

$$f_i(0) = f_i''(0) = 0, \quad f_1(1) + f_2(1) = 2, \quad f_i'(1) = 0.$$

When $\alpha = 0$, $f_2 \equiv 0$ and the two-dimensional case is recovered, whilst for $\alpha = 1$, $f_1 \equiv f_2$ corresponding to the above axisymmetric case. Wang and Watson discuss briefly numerical solutions of the equations for $f_i(\eta)$ for a range of values of R and α with $0 \leq \alpha \leq 1$. In particular they conclude that for $R > 0$, that is when fluid is squeezed out between the plates, the solution is unique. However for $R < 0$, corresponding to separation of the plates, dual solutions have been discovered for all values of α in the range $[0, 1]$. Table 5.1 shows values of A_0 for various values of α when $R = -1.0$.

5.5 Rotating-disk flows

5.5.1 Self-similar flows

Self-similar time-dependent solutions, analogous to those considered in the previous section, have been considered by several authors. For example Watson and Wang (1979) consider an infinite disk rotating with angular velocity

$$\Omega(t) = \frac{\Omega_0}{1 - kt}. \tag{5.55}$$

Then, with velocity components

$$v_r = \frac{\Omega_0 r}{1 - kt} f'(\eta), \quad v_\theta = \frac{\Omega_0 r}{1 - kt} g(\eta), \quad v_z = -\frac{2(\nu\Omega_0)^{1/2}}{(1 - kt)^{1/2}} f(\eta), \quad (5.56)$$

where $\eta = (\Omega_0/\nu)^{1/2} z/(1 - kt)^{1/2}$, the continuity equation (1.22) is satisfied identically and, with the pressure given from equation (1.21) as

$$\frac{p - p_0}{\rho} = \frac{\nu\Omega_0}{1 - kt}\{S\eta f(\eta) - 2f^2(\eta) - 2f'(\eta)\}, \quad S = \frac{k}{\Omega_0},$$

equations (1.19), (1.20) yield the following ordinary differential equations for $f(\eta)$, $g(\eta)$

$$f''' + 2ff'' - f'^2 + g^2 - S\left(f' + \tfrac{1}{2}\eta f''\right) = 0, \quad (5.57)$$

$$g'' + 2fg' - 2f'g - S\left(g + \tfrac{1}{2}\eta g'\right) = 0, \quad (5.58)$$

together with

$$f(0) = f'(0) = 0, \qquad g(0) = 1, \quad (5.59)$$

and

$$f'(\infty) = g(\infty) = 0. \quad (5.60)$$

Watson and Wang present solutions for a range of values of $S \leq 0$, indicating that no solutions exist for $S > 0$. The velocities therefore decay as t increases. The solutions have the properties that $g'(0) = 0$ for $S = S^* = -1.606699$, and $g'(0) \gtrless 0$ for $S \lessgtr S^*$. As the rate of decay of the disk rotation increases, that is as $|S|$ increases, the fluid close to the disk rotates more rapidly than the disk itself. For the special case $S = S^*$, for which $g'(0) = 0$, the torque on the disk vanishes, a case which Watson and Wang interpret as a situation corresponding to the decay of rotation of a free, massless disk in an infinite fluid. In figure 5.13 radial and azimuthal velocity profiles are shown for various values of S.

In the above example the fluid above the rotating disk is unbounded. Wang, Watson and Alexander (1991) consider a time-dependent rotating disk on which there is a liquid film of thickness $h(t)$. The angular velocity of the disk is again given by (5.55) with the velocity components as in equation (5.56). There is no radial pressure gradient so that f and g again satisfy (5.57), (5.58) and (5.59), but the conditions at the free surface now require careful consideration. At the surface of the film $v_z = dh/dt$ which indicates that $h(t)$ must be proportional to $(1 - kt)^{1/2}$. Writing

$$h(t) = \left(\frac{\nu}{\Omega_0}\right)^{1/2} (1 - kt)^{1/2} \beta, \quad (5.61)$$

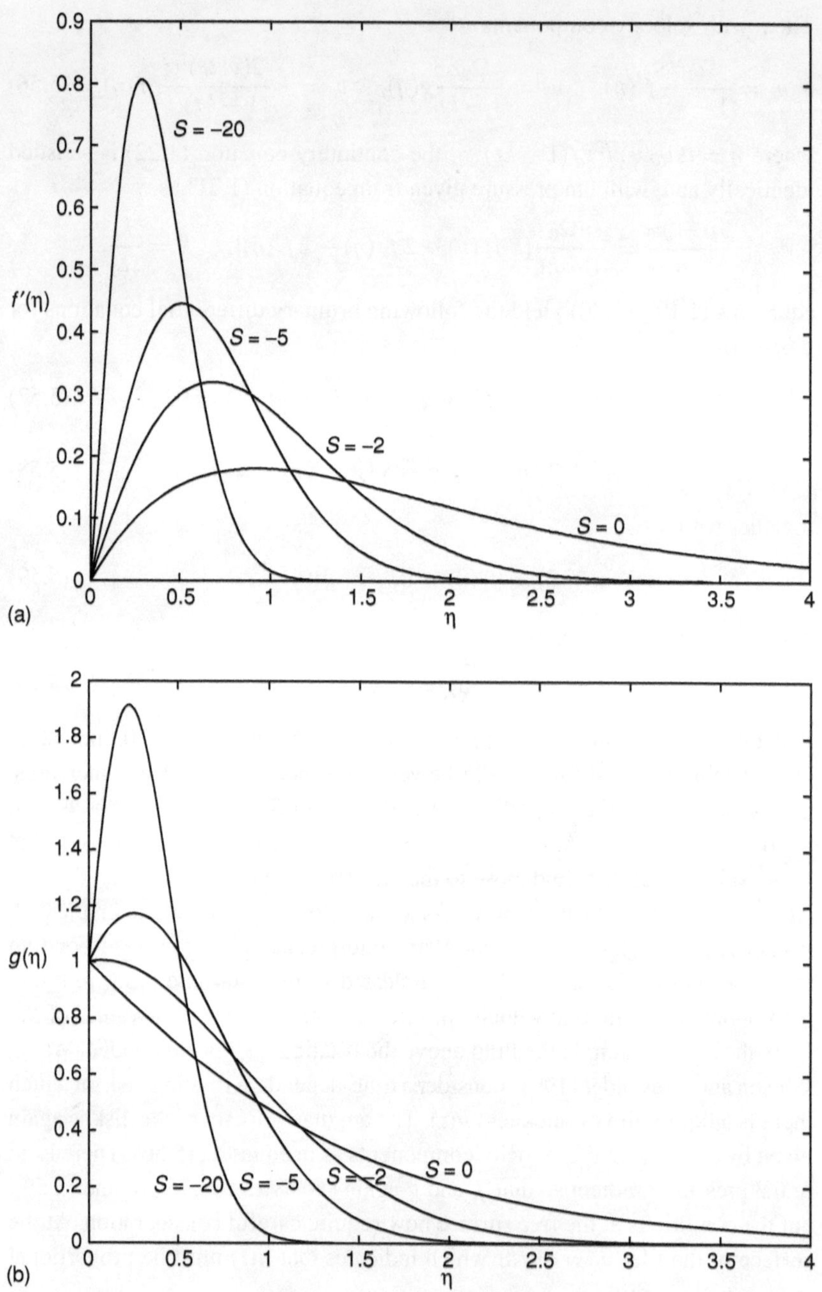

Figure 5.13 (a) The radial velocity profile $f'(\eta)$ for various values of S. (b) The azimuthal velocity profile $g(\eta)$ for various values of S.

the kinematic surface condition at $\eta = \beta$ is

$$f(\beta) = \frac{1}{4}\beta S \tag{5.62}$$

where, as before, $S = k/\Omega_0$. At the free surface the shear stress components vanish so that

$$f''(\beta) = g'(\beta) = 0. \tag{5.63}$$

As liquid is centrifuged radially outwards, the film thickness diminishes and, from (5.61), finally vanishes at $t = k^{-1}$. The unknowns $f(\eta)$, $g(\eta)$, β are determined from equations (5.57), (5.58) and the conditions (5.59), (5.62) and (5.63). Numerical solutions obtained by Wang *et al.* show that there are no solutions for $S > 0.588$ and that for $0 \leq S \leq 0.588$ two solutions exist. A summary of the dual solutions is shown in figure 5.14. For small values of β, f' is essentially parabolic with $g \approx 1$. But as β becomes large the liquid layer thickens, and indeed for S, $\beta^{-1} \rightarrow 0$ the solution approaches the steady von Kármán rotating disk solution, acknowledged by the intercepts for $f''(0)$, $g'(0)$ at $S = 0$ in figure 5.14(b). With p_0 the ambient pressure at $z = h$ the pressure in the liquid layer is given as

$$\frac{p - p_0}{\rho} = \frac{\nu\Omega}{1 - kt}\left\{ S\eta f(\eta) - 2f^2(\eta) - 2f'(\eta) + 2f'(\beta) - \frac{1}{8}\beta^2 S^2 \right\}.$$

Wang *et al.* have also discovered another branch of solutions for $-\infty < S < 0.0326$, but these have been rejected on physical grounds involving, as they do, negative radial velocity. Hamza and MacDonald (1984) also investigate the flow within a fluid layer of finite, but diminishing, thickness with an azimuthal velocity induced by not one, but two, rotating bounding surfaces. The flow may be interpreted as the squeeze flow of Wang (1976), discussed in section 5.4.2, upon which is superposed a rotation due to the rotation of one or both of the bounding disks. Adopting a notation in sympathy with the previously considered squeeze flows, the velocity components and pressure are written as

$$v_r = \frac{kr}{4(1 - kt)}f'(\eta), \quad v_\theta = \frac{\Omega_1 r}{(1 - kt)}g(\eta),$$

$$v_z = -\frac{kh_0}{2(1 - kt)^{1/2}}f(\eta), \quad \eta = \frac{z}{h_0(1 - kt)^{1/2}},$$

$$\frac{p - p_0}{\rho} = -\frac{A_0 k^2 r^2}{16(1 - kt)^2} - \frac{h_0^2 k^2}{8(1 - kt)}\left(\frac{4}{R}f'(\eta) - 2\eta f(\eta) + f^2(\eta)\right),$$

$$R = \frac{h_0^2 k}{\nu}.$$

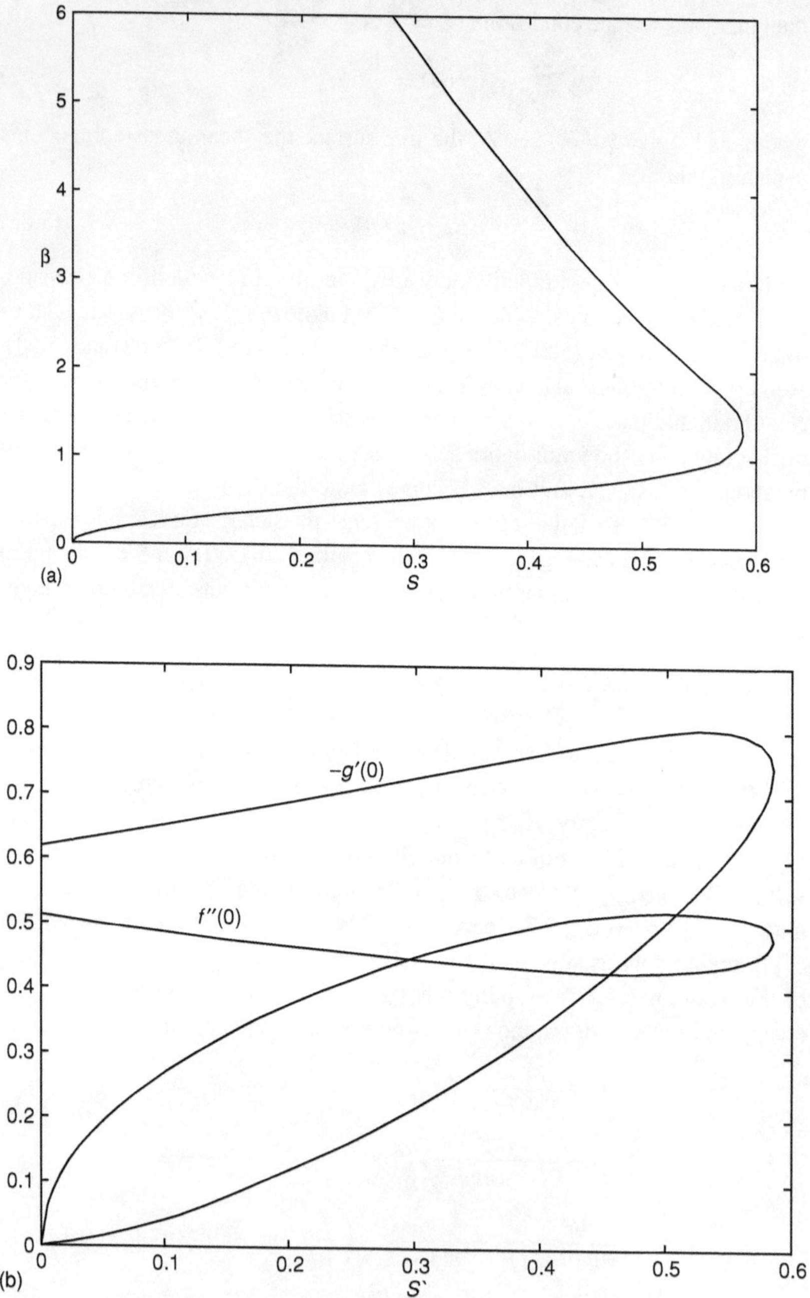

Figure 5.14 (a) Variation of the film thickness parameter β with S. (b) The radial and azimuthal shear-stress parameters $f''(0)$, $g'(0)$ as functions of S.

The continuity equation (1.22) is identically satisfied as is the normal momentum equation (1.21). The radial and azimuthal momentum equations will be satisfied provided that $f(\eta)$, $g(\eta)$ satisfy the ordinary differential equations

$$\frac{2}{R}f''' + ff'' - \frac{1}{2}f'^2 - 2f' - \eta f'' + \frac{8}{S^2}g^2 = -A_0, \qquad (5.64)$$

$$\frac{2}{R}g'' + fg' - f'g - 2g - \eta g' = 0, \qquad (5.65)$$

where $S = k\Omega_1$. The boundary conditions for equations (5.64), (5.65) will depend upon the relative rotation rates of the bounding disks. If one disk is at $z = 0$ rotating with angular velocity Ω_1, the other at $z = h_0(1 - kt)^{1/2}$, or $\eta = 1$, rotating with angular velocity Ω_2, then

$$f(0) = f'(0) = 0, \quad g(0) = 1; \quad f(1) = 1, \quad f'(1) = 0, \quad g(1) = \Omega_2/\Omega_1 = s. \qquad (5.66)$$

The solution of (5.64) to (5.66) depends upon the three dimensionless parameters R, S and s; solutions are presented by Hamza and MacDonald for various values of these. The cases $s = 0$ and $s = -1$ are of particular interest. In the former case $f' > 0$ so that there is radial outflow throughout the fluid layer. For fixed R, as S decreases f' assumes a boundary-layer character close to $\eta = 0$ with $f'_{\max}$ increasing. For the case of counter-rotating disks, $s = -1$, the flow character changes significantly for R fixed as S decreases. For sufficiently large S, $f' > 0$, and there is radial outflow throughout with symmetry about $\eta = \frac{1}{2}$. However, as S decreases, whilst fluid close to each rotating disk continues to be centrifuged radially outwards, eventually an *inward* radial flow develops in the central portion of the fluid layer, with symmetry again maintained about $\eta = \frac{1}{2}$.

5.5.2 Rotating disk in a counter-rotating fluid

In section 3.5 on steady rotating disk flows it was noted that if, in the notation of equation (3.43), $\lambda_1 = \lambda_2 = 0$, $\lambda_3 = -1$ then no solution of equations (3.41), (3.42) exists (McLeod (1971)). Bodonyi and Stewartson (1977) have pursued this case by considering an initial-value problem and integrating the governing equations forward in time. The investigation seeks the nature of the solution as time increases which cannot, of course, be a steady state. If the disk rotates with angular velocity Ω in its own plane then (3.39) is replaced by

$$v_r = r\Omega f_\eta(\eta, \tau), \quad v_\theta = r\Omega g(\eta, \tau), \quad v_z = -2(\nu\Omega)^{1/2}f(\eta, \tau), \qquad (5.67)$$

where $\eta = (\Omega/\nu)^{1/2}z$, $\tau = \Omega t$; and with the pressure given as

$$\frac{p - p_0}{\rho} = \frac{1}{2}\Omega^2 r^2 + 2\nu\Omega\left(\int_0^\eta f_\tau \, d\eta - f^2(\eta) - f'(\eta)\right),$$

the equations satisfied by $f(\eta, \tau)$, $g(\eta, \tau)$, namely the unsteady analogues of (3.41), (3.42), are

$$f_{\eta\tau} - f_{\eta\eta\eta} - 2ff_{\eta\eta} - f_\eta^2 - g^2 + 1 = 0, \tag{5.68}$$

$$g_\tau - g_{\eta\eta} - 2(fg_\eta - f_\eta g) = 0, \tag{5.69}$$

with

$$f(0, \tau) = f_\eta(0, \tau) = 0, \quad g(0, \tau) = 1; \qquad f_\eta(\infty, \tau) = 0, \quad g(\infty, \tau) = -1. \tag{5.70}$$

The conditions at infinity reflect the fact that far from the rotating disk the fluid is in solid-body rotation with angular velocity $-\Omega$. Bodonyi and Stewartson integrate equations (5.68), (5.69), together with conditions (5.70), forward in time starting from the initial state $f(\eta, 0) = f_\eta(\eta, 0) = 0$, $g(\eta, 0) = -1$ for all η. The solution for g quickly develops a non-monotonic state with values of $g > 1$ and $g < -1$. Moreover the maximum values of both f_η and $|g|$ develop explosively for values of $\tau > 2$, such that the integration could not be continued beyond $\tau = 2.25$. The nature of the singular 'blow-up' of the solution was inferred heuristically from the behaviour of $f(\infty, \tau)$. Figure 5.15 illustrates this and, in particular, indicates that as $\tau \to \tau_s$, $f(\infty, \tau) \sim C/(\tau_s - \tau)^2$ where C is a constant and the singular time $\tau_s \approx 2.365$. This behaviour provides a starting point for the asymptotic analysis carried out by Bodonyi and Stewartson who show that both the radial and azimuthal velocity components break down as $(\tau_s - \tau)^{-1}$, as the singular time is approached.

The solution (5.67) is of separable type. Grundy and McLaughlin (1999) have considered the behaviour of a separable three-dimensional solution of more general type. The flow takes place between the planes $z = \pm h_0$ and, in Cartesian co-ordinates, with

$$u = kxf_1(\eta, \tau), \quad v = kyf_2(\eta, \tau), \quad w = -kh_0g(\eta, \tau) \text{ where } \tau = kt, \eta = z/h_0,$$

the Navier–Stokes equations require

$$f_{i\eta\tau} - \frac{1}{R}f_{i\eta\eta\eta} - gf_{i\eta\eta} + (2f_i - g_\eta)f_{i\eta} = 0, \quad R = \frac{kh_0^2}{\nu}, \quad i = 1, 2;$$

$$f_1 + f_2 + g_\eta = 0. \tag{5.71}$$

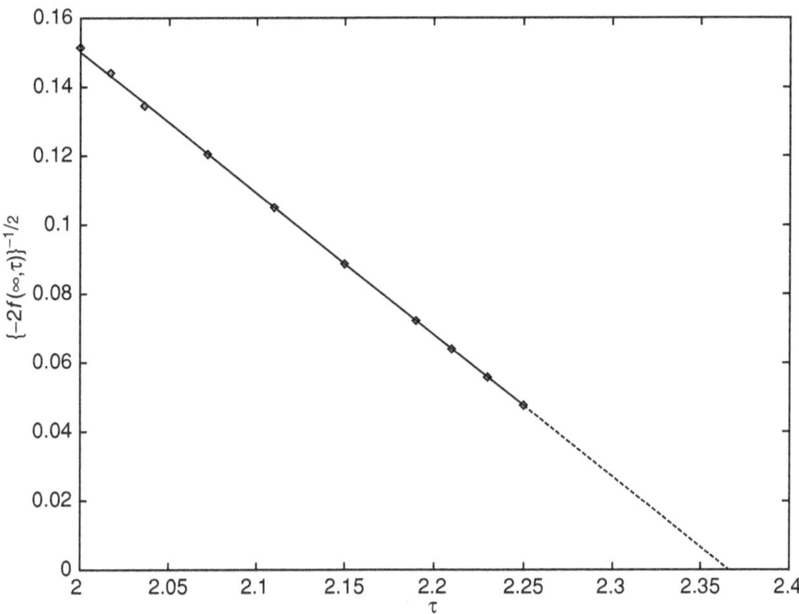

Figure 5.15 The linear variation of $\{-2f(\infty, \tau)\}^{-1/2}$ as $\tau \to \tau_s$ is demonstrated. The computed points are represented as ◇, with the broken line a linear extrapolation to τ_s.

Although the problem as posed by Grundy and McLaughlin has its origin in solar magnetohydrodynamics, it does represent a type of stagnation-point flow. The plane $z = 0$ is assumed to be a plane of symmetry so that

$$f_{1\eta} = f_{2\eta} = g = 0 \quad \text{at} \quad \eta = 0, \tag{5.72}$$

and at $z = h_0$ the boundary is supposed porous with prescribed stresses so that

$$f_{1\eta} = \alpha, \quad f_{2\eta} = \beta, \quad g = \gamma \quad \text{at} \quad \eta = 1. \tag{5.73}$$

The system of equations (5.71) is seventh order and so an additional condition is required to supplement those set out in (5.72), (5.73). This is taken as $f_2(0, \tau) = \delta f_1(0, \tau)$ which, in some sense, characterises the geometry of the stagnation-point flow on $z = 0$. The equations (5.71) are integrated forward in time from initial distributions of f_1, f_2 and g of polynomial type that are consistent with (5.72) and (5.73). In some cases, for example when $R = 0.5$, $\alpha = \beta = -15$, $\gamma = 1, \delta = 0.5$, a steady state solution is attained; for others, for example when $R = 5$, $\alpha = \beta = -15$, $\gamma = 1$, $\delta = 0.5$ the solution blows up at a finite time. In those cases where the solution does terminate in a singularity, the singular behaviour of u, v is as $(\tau_s - \tau)^{-1}$ and therefore similar to that exhibited in the problem addressed by Bodonyi and Stewartson (1977).

5.5.3 Non-axisymmetric flows

In section 3.5 attention has been drawn to the non-axisymmetric or eccentric flow between two parallel disks that rotate with the same angular velocity Ω but about different axes. Smith (1987) has included unsteady effects in such flows. In place of (3.51) the radial and azimuthal velocity components are written as

$$v_r = \Omega l\{f(\eta, \tau)\cos\theta + g(\eta, \tau)\sin\theta\},$$
$$v_\theta = \Omega r + \Omega l\{g(\eta, \tau)\cos\theta - f(\eta, \tau)\sin\theta\}, \tag{5.74}$$

with $v_z \equiv 0$ and the pressure given as

$$\frac{p - p_0}{\rho} = \frac{1}{2}\Omega^2 r^2 + \Omega^2 lr\{G(t)\cos\theta - F(t)\sin\theta\}. \tag{5.75}$$

The variables $\tau = \Omega t$, $\eta = z/h$ where h is the distance between the two disks; if only one disk is present then h is an arbitrary length. With the velocity components and pressure as in equations (5.74) and (5.75), equations (1.19) and (1.20) require

$$f_\tau - g - \frac{1}{R}f_{\eta\eta} = -G, \quad g_\tau + f - \frac{1}{R}g_{\eta\eta} = F, \quad \text{where again} \quad R = \frac{\Omega h^2}{\nu}. \tag{5.76}$$

The simplest situations arise when the upper disk is absent with fluid in the half-space $z \geq 0$ bounded by a single rotating disk at $z = 0$; for if the fluid at infinity rotates as a solid body with angular velocity Ω about the z-axis then $f, g \to 0$ as $\eta \to \infty$ and so $F = G \equiv 0$.

Consider first the case in which disk and fluid are in solid-body rotation about the common axis $r = 0$ for $\tau < 0$ and that the axis of rotation of the disk is moved impulsively from $r = 0$ to $r = l$, $\theta = \frac{1}{2}\pi$ at time $\tau = 0$. The boundary conditions for equations (5.76) are then

$$f = g = 0, \tau = 0, \eta > 0; f, g \to 0 \text{ as } \eta \to \infty; f = 1, g = 0, \eta = 0, \tau \geq 0. \tag{5.77}$$

Equations (5.76) subject to the conditions (5.77) may be solved using Laplace transform techniques. Thus the transforms $\bar{f}, \bar{g}$ of f, g are

$$\bar{f} = \frac{1}{2s}\left\{e^{-(s+i)^{1/2}\zeta} + e^{-(s-i)^{1/2}\zeta}\right\}, \quad \bar{g} = -\frac{i}{2s}\left\{e^{-(s+i)^{1/2}\zeta} - e^{-(s-i)^{1/2}\zeta}\right\},$$
$$\zeta = R^{1/2}\eta$$

which, upon inversion, give $f = \frac{1}{2}\mathrm{Re}(\phi)$, $g = \frac{1}{2}\mathrm{Im}(\phi)$ where

$$\phi = e^{-(1+i)\zeta/\sqrt{2}}\,\mathrm{erfc}\left(\frac{\zeta}{2\sqrt{\tau}} - \frac{1+i}{2\sqrt{2}}\sqrt{\tau}\right)$$
$$+ e^{(1+i)\zeta/\sqrt{2}}\,\mathrm{erfc}\left(\frac{\zeta}{2\sqrt{\tau}} + \frac{1+i}{2\sqrt{2}}\sqrt{\tau}\right).$$

As $\tau \to \infty$, $f \to e^{-\zeta/\sqrt{2}}\cos(\zeta/\sqrt{2})$, $g \to -e^{-\zeta/\sqrt{2}}\sin(\zeta/\sqrt{2})$ so that there is a final steady state in which

$$v_r = \Omega l\, e^{-\zeta/\sqrt{2}}\cos(\theta + \zeta/\sqrt{2}), \quad v_\theta = \Omega r - \Omega l e^{-\zeta/\sqrt{2}}\sin(\theta + \zeta/\sqrt{2}),$$

and each plane rotates with angular velocity Ω about an axis

$$r = l\exp\left\{-\left(\frac{R}{2}\right)^{1/2}\eta\right\}, \quad \theta = \frac{1}{2}\pi - \left(\frac{R}{2}\right)^{1/2}\eta.$$

In a further example the axis of rotation of the disk follows a circular orbit for $\tau > 0$ such that $f(0,\tau) = \cos\omega\tau$, $g(0,\tau) = -\sin\omega\tau$. The same technique may be used to determine f and g; in particular as $\tau \to \infty$ a periodic motion emerges with $f(\eta,\tau) = e^{-\lambda\zeta}\cos(\omega\tau + \lambda\zeta)$, $g(\eta,\tau) = -e^{-\lambda\zeta}\sin(\omega\tau + \lambda\zeta)$ where $\lambda = \{(1-\omega)/2\}^{1/2}$. A case of particular interest is when $\omega = 1$ so that the frequency of rotation of the fluid and that for the axis of the disk about the central axis are equal. For that case $f = \cos\tau\,\mathrm{erfc}(\zeta/2\sqrt{\tau})$, $g = -\sin\tau\,\mathrm{erfc}(\zeta/2\sqrt{\tau})$ and in the final state as $\tau \to \infty$

$$v_r = \Omega l\cos(\theta + \tau), \quad v_\theta = \Omega r - \Omega l\sin(\theta + \tau).$$

The velocity is independent of η; the rotation of the disk about its axis, and of its axis about the origin, combine to give eccentric solid-body rotation throughout the fluid. Smith observes that such a motion can always be induced by an appropriate periodic behaviour, period $2\pi/\Omega$, of the disk. This example in which the axis of rotation of the disk follows a periodic circular orbit has also been considered by Rao and Kasiviswanathan (1987).

The difficulties associated with the extension of the above to eccentric flow between two rotating disks have been discussed by Smith and are linked to the determination of $F(t)$, $G(t)$ in the expression (5.75) for the pressure. However, in the particularly important case

$$f(0,\tau) = -1, \quad g(0,\tau) = 0; \quad f(h,\tau) = 1, \quad g(h,\tau) = 0 \quad \text{for} \quad \tau \geq 0,$$

which yields the steady state solution of Abbott and Walters (1970) as $\tau \to \infty$, the mean position of the axis of rotation is $r = 0$ which is sufficient to ensure $F, G \equiv 0$. The solution may again be pursued by transform techniques, and although the transform functions $\bar{f}(\eta, s)$, $\bar{g}(\eta, s)$ are not readily inverted they do yield the steady state solution of Abbott and Walters as $\tau \to \infty$.

In section 3.5.1 the work of Hewitt *et al.* (1999) was introduced in which asymmetry was introduced into the rotating-disk solution even though the boundary conditions favoured an axisymmetric solution. The same superposed asymmetry had earlier been introduced by Hall, Balakumar and Papageorgiu (1992). The solution proposed by Hall *et al.* may be written as

$$v_r = \Omega r\{f'(\eta, \tau) + \phi(\eta, \tau)\cos 2\theta\}, \quad v_\theta = \Omega r\{g(\eta, \tau) - \phi(\eta, \tau)\sin 2\theta\},$$
$$v_z = -2(\nu\Omega)^{1/2}f(\eta, \tau), \quad \text{where} \quad \tau = \Omega t \quad \text{and} \quad \eta = (\Omega/\nu)^{1/2}z,$$

with the pressure given by

$$\frac{p - p_0}{\rho} = \frac{1}{2}\lambda\Omega^2 r^2 + \nu\Omega P(\eta, \tau) + \Omega^2 r^2 Q(\tau)\cos 2\theta.$$

The continuity equation is satisfied identically and the momentum equations (1.19), (1.20), (1.21) give, for f, g, ϕ and P

$$f_{\eta\tau} - f_{\eta\eta\eta} - 2ff_{\eta\eta} + f_\eta^2 + \phi^2 - g^2 + \lambda = 0, \tag{5.78}$$

$$g_\tau - g_{\eta\eta} - 2fg_\eta + 2f_\eta g = 0, \tag{5.79}$$

$$\phi_\tau - \phi_{\eta\eta} - 2f\phi_\eta + 2f_\eta\phi + 2Q = 0, \tag{5.80}$$

$$P_\eta = 2f_\tau - 2f_{\eta\eta} - 4ff_\eta. \tag{5.81}$$

Whilst the main burden of their investigation centred upon unsteady motion, Hall *et al.* considered steady solutions of equations (5.78) to (5.80). The boundary conditions for these equations are

$$f = f_\eta = \phi = 0, \quad g = 1 \quad \text{at} \quad \eta = 0; \quad f_\eta, g \to 0, \quad \phi \to \gamma \quad \text{as} \quad \eta \to \infty \tag{5.82}$$

so that $\lambda = -\gamma^2$. The flow as $\eta \to \infty$ is, then, of a stagnation-point flow type. By contrast Hewitt *et al.* have $\gamma = 0$, but $g(\infty) = s \neq 0$ so that as $\eta \to \infty$ the flow is a solid-body rotation. In their numerical investigation of the resulting non-linear ordinary differential equations, Hall *et al.* reveal that for each value of γ there is a continuum of solutions arising from the decision to retain all solutions that decay exponentially and not rejecting those that decay most slowly. An illustration of the diverse solutions obtained is given in figure 5.16.

For unsteady flow attention is focused on the solution of (5.78) to (5.80) with γ and hence λ and Q equal to zero. The motivation is, in effect, to investigate the finite-amplitude instability of von Kármán's classical rotating-disk solution although, of course, the superposed finite-amplitude disturbances are restricted

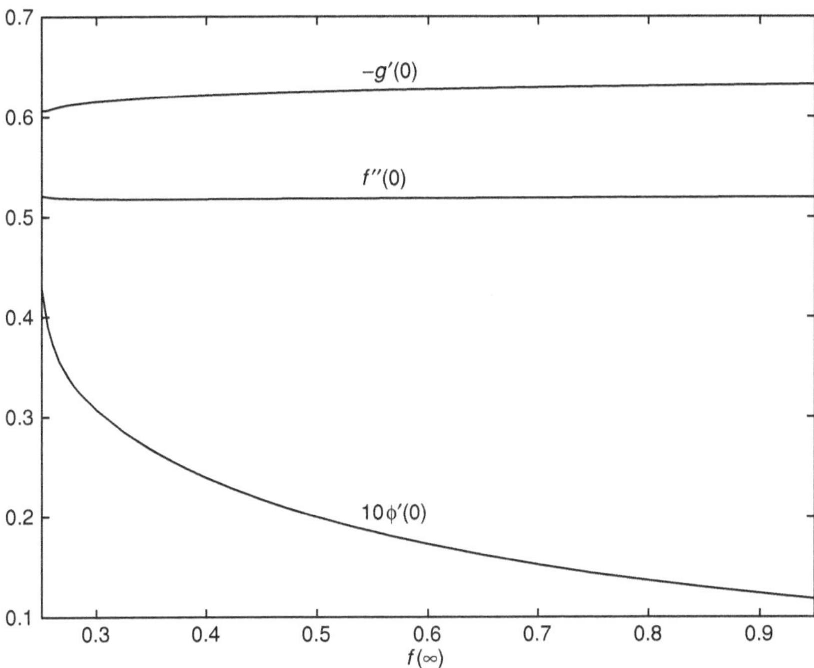

Figure 5.16 The shear-stress parameters $f''(0)$, $g'(0)$ and $\phi'(0)$ plotted as functions of $f(\infty)$.

to azimuthal wavenumber ± 2. The approach adopted is to integrate the equations forward in time. To enable this, initial conditions at $\tau = 0$ are required for f', g and ϕ; these are set as

$$f'(\eta, 0) = g(\eta, 0) = 0, \quad \phi(\eta, 0) = \delta\eta \, e^{-(\eta-2)^2}, \tag{5.83}$$

and the calculations are carried out for values of $\delta = 0.35$, 0.45, 0.55, 0.65. For the two smaller values of δ the solution decays to the classical von Kármán solution. However, for the two larger values the solution terminates in a singularity at a finite time τ_s. In this 'blow-up' of the solution both v_r and v_θ exhibit behaviour $(\tau_s - \tau)^{-1}$ as $\tau \to \tau_s^-$, and so this is of the same type as observed in the blow-up of solutions encountered in the previous section.

5.5.4 An Ekman flow

In section 3.6 the classical steady Ekman (1905) solution is set out. In fact Ekman also considered an unsteady problem based on the impulsive start of a uniform shear in the rotating system. Rott and Lewellen (1967) have also considered an oscillatory type of Ekman flow. If the boundary rotates with angular velocity Ω about an axis $\{x_c(t), y_c(t)\}$ then, in Cartesian co-ordidates, the velocity components at the plate are

$$u_w = -\Omega(y - y_c) + \frac{dx_c}{dt}, \quad v_w = \Omega(x - x_c) + \frac{dy_c}{dt},$$

and so, if

$$u = -\Omega y + U h_1(z, t), \quad v = \Omega x + U h_2(z, t)$$

then

$$U\{h_1(0, t) + ih_2(0, t)\} = -i\Omega\chi + \frac{d\chi}{dt} \quad \text{where} \quad \chi = x_c + iy_c.$$

If a frame of reference rotating with angular velocity ω is introduced and $h(z, t)$ such that $h_1 + ih_2 = he^{i\omega t}$ then the real and imaginary parts of h are the Cartesian velocity components in this rotating frame. The equation satisfied by h is

$$\frac{\partial h}{\partial t} + i(\Omega + \omega)h = \nu\frac{\partial^2 h}{\partial z^2}, \tag{5.84}$$

with

$$h(0, t) = \frac{1}{U}\left(-i\Omega\chi + \frac{d\chi}{dt}\right) e^{-i\omega t}, \quad h \to 0 \quad \text{as} \quad z \to \infty.$$

If the centre of the boundary oscillates with frequency $\lambda/2\pi$ in this rotating frame then $\chi = l\, e^{i\omega t} \cos \lambda t$, and $h(0, t)$ becomes

$$h(0, t) = \frac{il}{2U}\{(\omega + \lambda - \Omega)\, e^{i\lambda t} + (\omega - \lambda - \Omega)\, e^{-i\lambda t}\}. \tag{5.85}$$

Defining quantities $\Lambda_\pm = \omega + \Omega \pm \lambda$, and assuming for simplicity that $\Lambda_\pm > 0$, the solution of (5.84) subject to (5.85) that decays as $z \to \infty$ is

$$h(z, t) = \frac{l}{U}(\omega + \lambda - \Omega)\, e^{-(\Lambda_+/2\nu)^{1/2}z} \sin\left\{\left(\frac{\Lambda_+}{2\nu}\right)^{1/2} z - \lambda t\right\}$$

$$+ \frac{l}{2U}(\omega - \lambda - \Omega)\, e^{-(\Lambda_-/2\nu)^{1/2}z} \sin\left\{\left(\frac{\Lambda_-}{2\nu}\right)^{1/2} z - \lambda t\right\}.$$

For $\Omega = 0$ Stokes' solution considered in section 4.1 is recovered, albeit with a rotating plane of oscillation.

5.6 Vortex motion

5.6.1 Single-cell vortices

In section 3.7.2 steady vortex flows in an unbounded fluid domain were considered of both single-cell (Burgers (1948)) and two-cell (Sullivan (1959)) structures. For a single-cell structure the radial velocity v_r does not change sign, whilst for a multi-cell structure v_r will change sign, once for the two-cell vortex of Sullivan.

For unsteady flow the classical solution of an isolated line vortex in an otherwise undisturbed fluid was first discussed by Oseen (1911) with

$$v_r = 0, \quad v_\theta = \frac{\Gamma}{2\pi r} g(r, t), \quad v_z = 0, \tag{5.86}$$

which may be compared with (3.63), (3.64). The equation for g is, from (1.20),

$$\frac{\partial g}{\partial t} = \nu \left(\frac{\partial^2 g}{\partial r^2} - \frac{1}{r} \frac{\partial g}{\partial r} \right),$$

and a solution which is regular at $r = 0$ and for which $g \to 1$ as $r \to \infty$ has

$$g(r, t) = 1 - e^{-r^2/4\nu t}, \tag{5.87}$$

which may be compared with the solution (3.65). This solution (5.87) also has the property that $g \to 1$ as $t \to 0$ so that $v_\theta \sim \Gamma/2\pi r$, the potential vortex, in that limit. Initially, then, all the vorticity is concentrated along the axis $r = 0$. As t increases v_θ decreases at a given radial distance r, but as $r \to \infty$ the potential vortex solution is recovered with circulation Γ. The circulation about a closed circuit at large radial distances is, of course, unchanged since the total vorticity within it is unchanged. As $r \to 0$, at a fixed t, $v_\theta \sim \Gamma r/8\pi \nu t$, which shows that in the viscous core of the vortex the fluid is in solid-body rotation, with a diminishing angular velocity as t increases. In figure 5.17 the development of the flow is depicted.

Oseen's solution may be generalised by writing, instead of (5.86),

$$v_r = -\gamma(t)r, \quad v_\theta = \frac{\Gamma}{2\pi r} g(r, t), \quad v_z = 2\gamma(t)z \tag{5.88}$$

so that, from (1.20), g now satisfies

$$\frac{\partial g}{\partial t} - \gamma(t)r \frac{\partial g}{\partial r} = \nu \left(\frac{\partial^2 g}{\partial r^2} - \frac{1}{r} \frac{\partial g}{\partial r} \right). \tag{5.89}$$

This equation admits a self-similar solution with

$$\eta = \frac{r}{\phi(t)}, \tag{5.90}$$

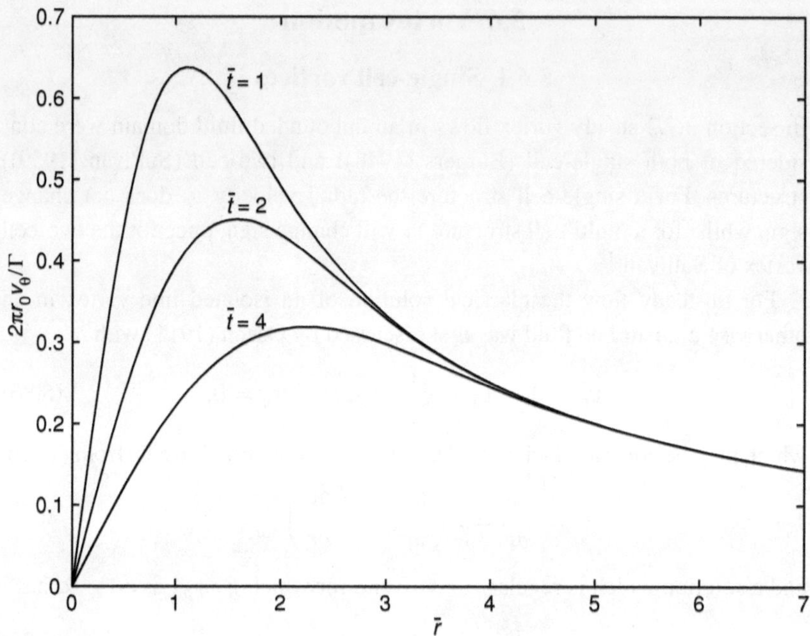

Figure 5.17 Velocity profiles in the classical Oseen (1911) vortex shown as a function of $\bar{r} = r/r_0$ for various values of $\bar{t} = 4\nu t/r_0^2$, where r_0 is an arbitrary length.

so that

$$-(\phi\dot{\phi} + \gamma\phi^2)\eta\frac{\partial g}{\partial \eta} = \nu\left(\frac{\partial^2 g}{\partial \eta^2} - \frac{1}{\eta}\frac{\partial g}{\partial \eta}\right) \tag{5.91}$$

provided that

$$\phi\dot{\phi} + \gamma\phi^2 = a, \quad \text{constant,} \tag{5.92}$$

in which case the solution for g is

$$g(r, t) = 1 - e^{-a\eta^2/2\nu} = 1 - e^{-ar^2/2\nu\phi^2}, \tag{5.93}$$

which may be compared with (5.87). Using (5.88) and (5.92) in equations (1.19) and (1.21) yields, for the pressure,

$$\frac{p - p_0}{\rho} = \frac{1}{2}\left\{(\dot{\gamma} - \gamma^2)r^2 - 2(\dot{\gamma} + 2\gamma^2)z^2\right.$$

$$\left. - \frac{\Gamma^2 a}{8\pi^2\nu\phi^2}\int_{ar^2/2\nu\phi^2}^{\infty}\left(\frac{1 - e^{-s}}{s}\right)^2 \,\mathrm{d}s\right\}. \tag{5.94}$$

The solution is completed when ϕ is determined from (5.92). This first-order equation for ϕ^2 has solution

$$\phi^2 = 2a \exp\left(-2\int_0^t \gamma(s)\,\mathrm{d}s\right) \int_c^t \exp\left(2\int_0^u \gamma(s)\,\mathrm{d}s\right)\,\mathrm{d}u, \qquad (5.95)$$

as in Rott (1958), where c is arbitrary.

Special cases of the unsteady stagnation-point flow represented by v_r and v_z in (5.88) may be considered. If $\gamma \equiv 0$ then $\phi = (2at)^{1/2}$ and Oseen's solution (5.87) is immediately recovered. The case $\gamma = k$, constant, has been discussed by Rott. This corresponds to the steady stagnation flow (3.63) and, indeed, if in (5.95) $c = -\infty$, $\phi^2 = a/k$ Burgers' solution (3.65) is recovered. For arbitrary c equation (5.95) yields

$$\phi^2 = \frac{a}{k}(1 + \beta\, \mathrm{e}^{-2kt}), \qquad (5.96)$$

where β is arbitrary, so that the solution decays to Burgers' solution as $t \to \infty$. A case which exhibits singular behaviour at a finite time has been studied in detail by Moffatt (2000). With $\gamma = (t_s - t)^{-1}$ the corresponding solution of (5.95), with $c = -\infty$, is $\phi = \{2a(t_s - t)\}^{1/2}$. The flow develops from $v_\theta = 0$ at $t = -\infty$ to the potential vortex $v_\theta = \Gamma/2\pi r$ as $t \to t_s$, when the stagnation velocity components v_r, v_z become unbounded. Amongst the other possibilities, and they are endless, consider the following algebraically decaying stagnation flow for which $\gamma = 1/2(t_0 + t)$ so that, from (5.92), $\phi = a^{1/2}(t_0 + t)^{1/2}$ and hence

$$v_r = -\frac{r}{2(t_0 + t)}, \qquad v_\theta = \frac{\Gamma}{2\pi r}\left(1 - \mathrm{e}^{-r^2/2\nu(t_0+t)}\right), \qquad v_z = \frac{z}{(t_0 + t)}.$$

Then, as $t \to \infty$

$$v_r, \quad v_z \to 0, \qquad v_\theta \sim \frac{\Gamma}{2\pi r}\left(1 - \mathrm{e}^{-r^2/2\nu t}\right) = \frac{\Gamma}{2\pi r}\left(1 - \mathrm{e}^{-r^2/4\nu_e t}\right)$$

$$(5.97)$$

where $\nu_e = \nu/2$. The solution (5.97) incorporates the Oseen solution (5.87) but with an effective kinematic viscosity ν_e reflecting the different dynamical routes through which the flows have developed.

In the above examples the only component of the vorticity vector $\boldsymbol{\omega}$ is the axial component $\omega_z = (\Gamma/2\pi r)\,\partial g/\partial r$. An extension retains v_r, v_θ as in (5.88) but has $v_z = 2\gamma(t)z + W(r, t)$ so that there is, in addition to the axial component of vorticity, an azimuthal component $\omega_\theta = -\partial W/\partial r$. The equation (5.91) for $g(r, t)$ is unchanged with the choice (5.95) for $\phi(t)$, and with the pressure as in equation (5.94) the equation satisfied by $W(r, t)$ is, from (1.21),

$$\frac{\partial W}{\partial t} - \gamma r \frac{\partial W}{\partial r} + 2\gamma W = \nu \left(\frac{\partial^2 W}{\partial r^2} + \frac{1}{r}\frac{\partial W}{\partial r}\right). \qquad (5.98)$$

Introducing the similarity variable (5.90) and employing the relationship (5.92) gives

$$\phi^2 \left(\frac{\partial W}{\partial t} + 2\gamma W \right) - a\eta \frac{\partial W}{\partial \eta} = v \left(\frac{\partial^2 W}{\partial \eta^2} + \frac{1}{\eta} \frac{\partial W}{\partial \eta} \right). \qquad (5.99)$$

A separable solution of this equation with

$$W = f(t)\, e^{-a\eta^2/2v}$$

gives, for $f(t)$,

$$\dot{f} + 2\left(\gamma + \frac{a}{\phi^2} \right) f = 0 \qquad (5.100)$$

with formal solution

$$f(t) = C \exp\left[-2 \int_0^t \left\{ \gamma(s) + \frac{a}{\phi^2(s)} \right\} ds \right].$$

The special case $\gamma(t) = k$, constant, for which ϕ is as in (5.96), has been considered by Gibbon, Fokas and Doering (1999) with the solution of (5.100) as

$$f = \frac{e^{-4kt}}{\phi^2} = \frac{k}{a}(e^{4kt} + \beta\, e^{2kt})^{-1}.$$

So that for this special case $W(r,t)$ may be written as

$$W = (e^{4kt} + \beta\, e^{2kt})^{-1} \exp\{-kr^2(1 + \beta\, e^{-2kt})^{-1}/2v\}.$$

As $t \to \infty$, $W \to 0$, $\phi^2 \to a/k$ and this solution again approaches the steady solution of Burgers as in equations (3.63) to (3.65).

5.6.2 Multi-cell vortices

Bellamy-Knights (1970) has introduced the analogue of the two-cell solution of Sullivan (1959), set out in equations (3.66), (3.67). In terms of the similarity variable $\eta = r/t^{1/2}$, Bellamy-Knights' solution may be written as

$$v_r = \left(k - \frac{1}{2} \right) t^{-1/2} \left\{ -\eta + \frac{6v}{k\eta}(1 - e^{-k\eta^2/2v}) \right\},$$

$$v_\theta = \frac{\Gamma}{2\pi r} \frac{g(\eta)}{g(\infty)}, \qquad v_z = (2k-1)zt^{-1}(1 - 3\,e^{-k\eta^2/2v}),$$

$$(5.101)$$

where

$$g(\eta) = \int_0^{\eta^2/4v} \exp\left\{ -u + 3\left(1 - \frac{1}{2k} \right) \int_0^u \left(\frac{1 - e^{-s}}{s} \right) ds \right\} du,$$

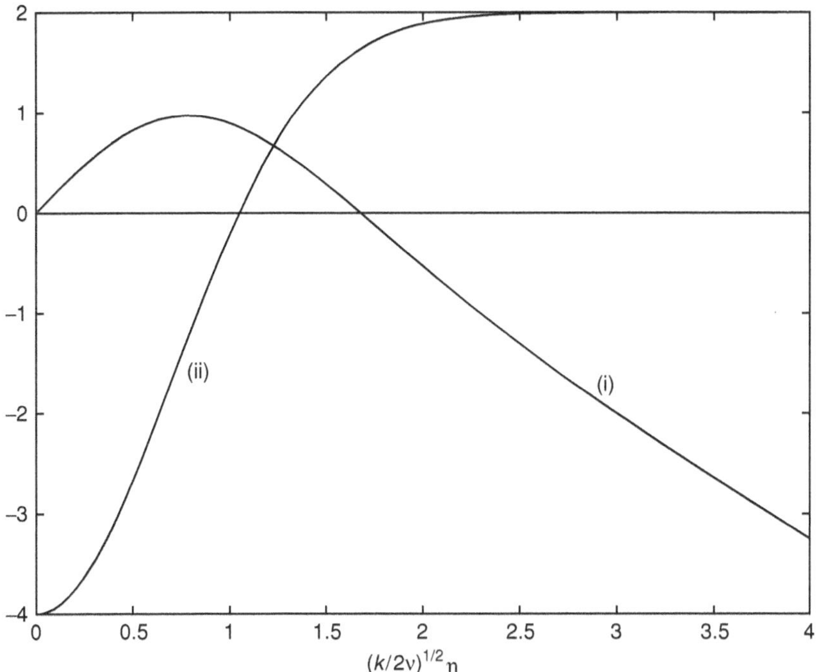

Figure 5.18 Velocity profiles in the two-cell vortex of Bellamy-Knights (1970).
(i) $v_r(kt/2v)^{1/2}/(k - \frac{1}{2})$, (ii) $v_z(t/z)/(k - \frac{1}{2})$.

which may be compared with (3.66), (3.67). The corresponding expression for
the pressure is

$$\frac{p - p_0}{\rho} = -\frac{(2k - 1)}{8t^2}\{8(k - 1)z^2 + (2k + 1)r^2\}$$

$$+ \frac{9v^2(2k - 1)^2}{2k^2}\frac{1}{t\eta^2}(1 - e^{-k\eta^2/2v})^2 + \int_0^\eta \frac{v_\theta^2}{s}\mathrm{d}s.$$

Whilst only positive values of k are permitted if the solution is to remain bounded
as $\eta \to \infty$, the character of the solution depends upon $k \gtrless \frac{1}{2}$ in the sense that
there is radial inflow far from the axis if $k > \frac{1}{2}$, outflow if $k < \frac{1}{2}$. In figure
5.18 the radial and axial velocity components are shown. As k increases, the
diameter of the inner cell, defined as the cylinder at which $v_r = 0$, decreases
due to the enhanced radial inflow. For all values of $k > \frac{1}{2}$ this bounding surface
between the two cells contracts radially at a rate $\propto (v/t)^{1/2}$. Stream surfaces
$z = z_0| - k\eta^2/2v + 3(1 - e^{-k\eta^2/2v})|^{-1}$, where z_0 is an arbitrary constant, are
shown in figure 5.19.

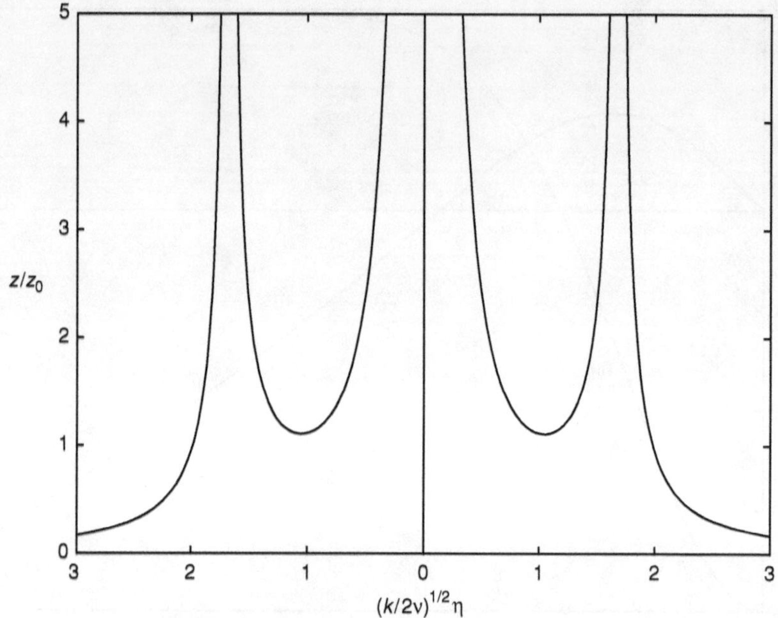

Figure 5.19 Stream surfaces of the two-cell vortex.

In a subsequent paper Bellamy-Knights (1971) has explored the possibility
of extending the range of the multi-cell solutions set out in equation (5.101).
Bellamy-Knights seeks solutions which have behaviour as $r \to \infty$ similar to
that of the velocity components in equation (5.101) so that

$$v_r \sim -\frac{\gamma r}{2t}, \quad v_\theta \sim \frac{\Gamma}{2\pi r}, \quad v_z \sim \frac{\gamma z}{t} \quad \text{as} \quad r \to \infty. \qquad (5.102)$$

A self-similar solution is available with

$$v_r = -\frac{2\nu}{r} f(\eta), \quad v_\theta = \frac{\Gamma}{2\pi r} \frac{g(\eta)}{g(\infty)}, \quad v_z = \frac{z}{t} f'(\eta) \quad \text{where now } \eta = \frac{r^2}{4\nu t}.$$

With the pressure given by

$$\frac{p - p_0}{\rho} = -\frac{1}{2}\gamma(\gamma - 1)\left(\frac{z}{t}\right)^2 - \frac{\nu}{t}\left(f(\eta) + f'(\eta) + \frac{f^2(\eta)}{2\eta}\right) + \frac{1}{2}\int_0^\eta \frac{v_\theta^2}{\eta}\, d\eta,$$

equations (1.21) and (1.20) yield, respectively, equations for $f(\eta)$ and $g(\eta)$ as

$$(\eta f'' + \eta f')' + ff'' - f'^2 + \gamma(\gamma - 1) = 0, \qquad (5.103)$$

and

$$\eta g'' + \eta g' + fg' = 0. \qquad (5.104)$$

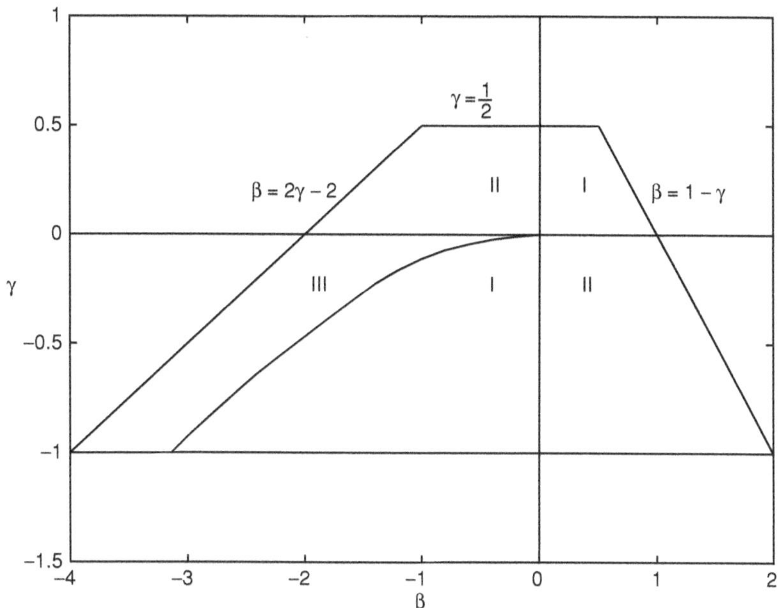

Figure 5.20 Regions of (β, γ)-space in which Bellamy-Knights (1971) finds: I, one-cell solutions; II, two-cell solutions; III, three-cell solutions.

The boundary conditions for equation (5.104) require $g(0) = 0$ if $v_\theta = 0$ at $r = 0$, and $g(\infty) = 1$ from (5.102). For equation (5.103), similarly, $f(0) = 0$ and $f'(\infty) = \gamma$. Bellamy-Knights also requires $f'''(0)$ to be finite at $\eta = 0$ so that $f''(0) = (\beta - \gamma)(\beta + \gamma - 1)$ where $f'(0) = \beta$ with the constant β unknown *a priori*. In the computational scheme the conditions adopted for (5.103) are

$$f(0) = 0, \quad f'(0) = \beta, \quad f''(0) = (\beta - \gamma)(\beta + \gamma - 1), \tag{5.105}$$

where a shooting method is adopted such that for each value of γ the value of β is chosen so that $f'(\infty) = \gamma$.

There are two exact solutions of (5.103), (5.104), (5.105). The first has $\gamma = \beta$ so that $f''(0) = 0$ and $f = \gamma\eta$ with $g = 1 - \exp\{-(1 + \gamma)\eta\}$. This is an example of the one-cell vortices in equation (5.88) with $v_r \propto r/t$. Another exact solution is available when $\gamma = -\beta/2$. In that case the solution of (5.103) is $f = \gamma\eta - \{3\gamma/(1 + \gamma)\}[1 - \exp\{-(1 + \gamma)\eta\}]$, which corresponds to the two-cell solution of equation (5.101) when $\gamma = 2k - 1$. A numerical investigation of (5.103), (5.105) has yielded further one- and two-cell solutions and, in addition, three-cell solutions. The regions of (β, γ)-parameter space in which these single- and multi-cell solutions have been found are shown in figure 5.20.

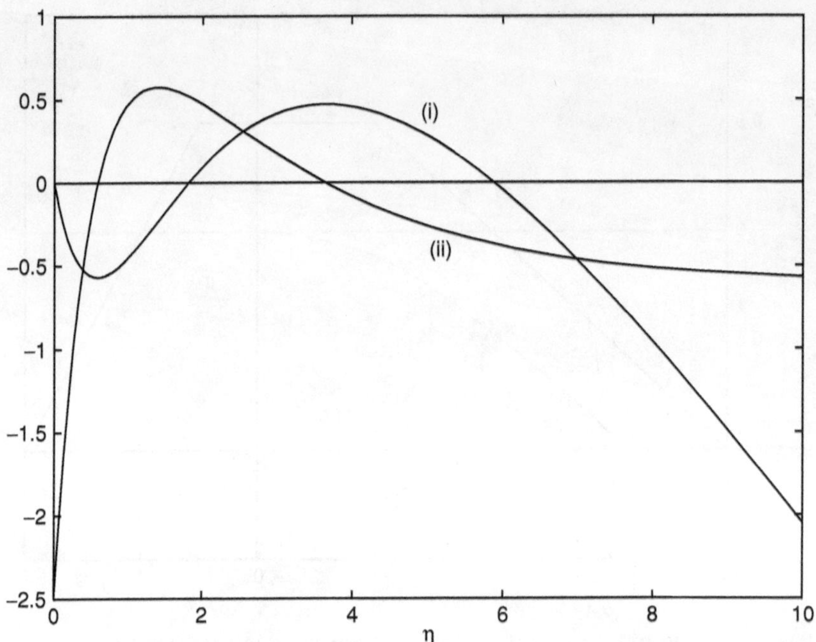

Figure 5.21 Velocity profiles for a three-cell vortex with $\beta = -2.5, \gamma = -0.6$. (i) $f(\eta) = -rv_r/2v$, (ii) $f'(\eta) = tv_z/z$, where $\eta = r^2/4vt$.

Apart from the analytic one- and two-cell solutions which are available for all $\gamma > -1$ on $\gamma = \beta, \gamma = -\beta/2$ respectively, other solutions have been found as follows (see figure 5.20). For $0 < \gamma < \frac{1}{2}$ solutions were obtained within the range $2\gamma - 2 < \beta < 1 - \gamma$; these are all of one-cell type for $\beta/\gamma > 0$ and two-cell type for $\beta/\gamma < 0$. For $-1 < \gamma < 0$ solutions were obtained for all β in the range $2\gamma - 2 < \beta < 1 - \gamma$; when $\beta > \beta_1(\gamma)$, where the boundary $\beta_1(\gamma)$ is determined numerically, the solutions are again one-cell for $\beta/\gamma > 0$ and two-cell for $\beta/\gamma < 0$. In the range $2\gamma - 2 < \beta < \beta_1(\gamma)$ three-cell solutions were found by Bellamy-Knights. These solutions, with $\gamma < 0$, all have the property that there is radial outflow from the vortex at large radial distances. By contrast it may be noted that Donaldson and Sullivan (1960) have established that only one- and two-cell solutions of the Burgers (1948) and Sullivan (1959) type exist for steady flow. In figure 5.21 radial and axial velocity profiles are shown for $\beta = -2.5, \gamma = -0.6$. The cylindrical cell boundaries are given by $v_r = 0$, and these increase again at a rate $\propto (v/t)^{1/2}$. The corresponding stream surfaces $z = z_0/|f(\eta)|$ are shown in figure 5.22.

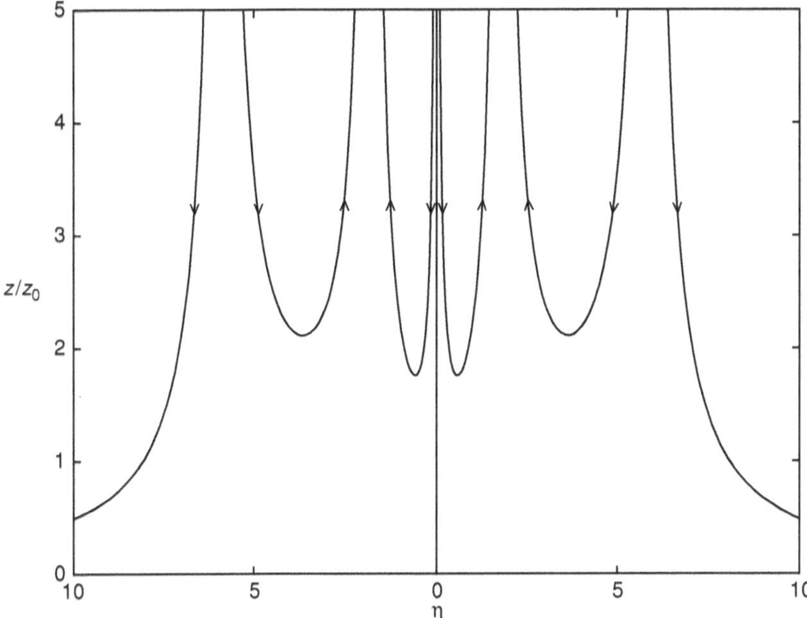

Figure 5.22 Stream surfaces corresponding to the profiles of figure 5.21.

For all of these multi-cell solutions $g(\eta) > 0$, increasing monotonically from zero to the unit asymptote so that the velocity distribution $v_\theta(r, t)$, at successive times, is qualitatively similar to the Oseen solution shown in figure 5.16.

5.6.3 The influence of boundaries

For the cellular vortex solutions, for example (5.88), (5.101), considered above all have the property that $v_z = 0$ at $z = 0$. The plane $z = 0$ may, then, be considered to be a solid boundary except, of course, the no-slip condition is not satisfied at it. One consequence, as in the steady-flow case, is that the strain rate and circulation at large distances are independent; see, for example, equation (5.101). Hatton (1975) introduces a no-slip boundary at $z = 0$, which couples the swirling and induced secondary motion, but finds that no exact solution is available corresponding to a potential vortex in the far field. Instead he proposes that at large distances from the boundary

$$v_r \sim -\frac{\gamma r}{2t}, \quad v_\theta \sim \frac{\Omega r}{4t}, \quad v_z \sim \frac{\gamma z}{t}, \tag{5.106}$$

which may be compared with (5.102). This corresponds to a flow established over the boundary which is in solid-body rotation with instantaneous angular velocity $\Omega/4t$ together with a radial and axial flow. Such solid-body rotation is a feature of the flow in the core of all the vortices considered above. Substituting (5.106) into equation (1.20) shows that it is necessary for $\gamma = -1$ so that in the core, away from the boundary, there is always a radial outflow and axial downflow.

With the velocity components and the pressure written as

$$v_r = \frac{r}{2t}f'(\eta), \quad v_\theta = \frac{\Omega r}{4t}g(\eta), \quad v_z = -2\left(\frac{\nu}{t}\right)^{1/2}f(\eta),$$

$$\frac{p - p_0}{\rho} = \frac{1}{32}(4 + \Omega^2)\left(\frac{r}{t}\right)^2 - \frac{2\nu}{t}\left\{f^2(\eta) + \frac{1}{2}f'(\eta) + \eta f(\eta)\right\},$$

where now $\eta = z/2(\nu t)^{1/2}$. Equations (1.19), (1.20) then give, as equations for f and g

$$f''' + 4ff'' - 2f'^2 + 4f' + 2\eta f'' + \tfrac{1}{2}\Omega^2 g^2 = \tfrac{1}{2}(4 + \Omega^2),$$

$$f(0) = f'(0) = 0, \quad f'(\infty) = 1,$$

and

$$g'' + 4(fg' - f'g) + 2\eta g' + 4g = 0,$$

with

$$g(0) = 0, \quad g(\infty) = 1.$$

The conditions at $\eta = \infty$ are determined by (5.106).

These equations for f and g have been integrated for a wide range of values of Ω. In all cases the behaviour of $g(\eta)$ is unexceptional with a monotonic increase to its asymptote. The behaviour of $f(\eta)$ is less straightforward. In figure 5.23 $f''(0)$ is shown as a function of Ω^2. For $\Omega^2 > 5$, $f''(0) < 0$ which means that there is a region close to the boundary, increasing with Ω^2, within which f, $f' < 0$ so that $v_r < 0$, $v_z > 0$, that is, radial inflow and axial upflow, before radial outflow and axial downflow are recovered in the far field. This implies the presence of a stagnation point on the axis. In figure 5.24 the location of this point is shown, together with the elevation at which v_r changes sign.

Hatton's work was motivated by the observed behaviour of tornadoes. There is visual evidence that close to the ground within the core of a vortex tornado there may be upflow or downflow with, in the former case, the upflow terminated by a stagnation point. The analysis of Hatton, based on the simple model that

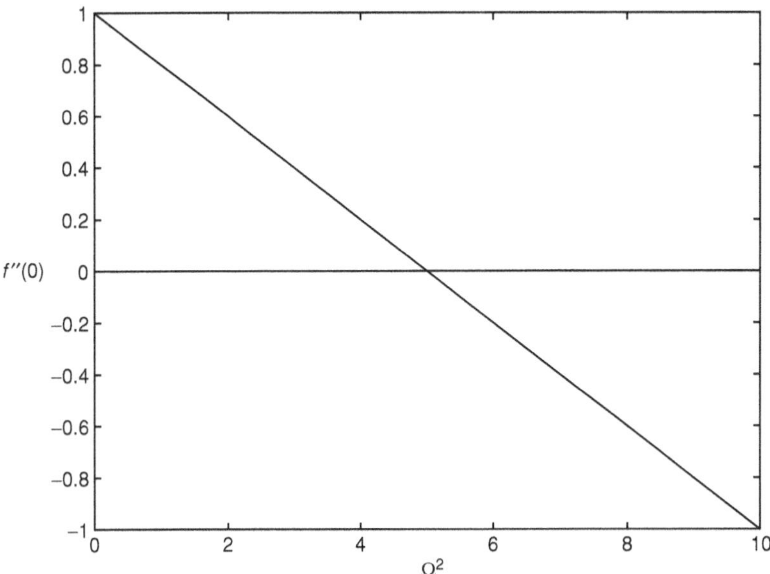

Figure 5.23 The shear-stress parameter $f''(0)$ as a function of Ω^2.

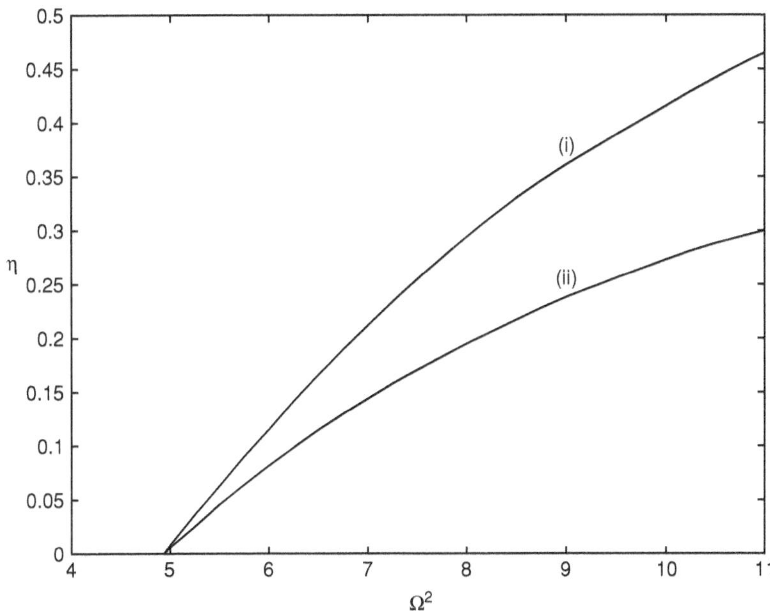

Figure 5.24 The variation with Ω^2 of: (i) the height of the stagnation point, (ii) the height at which the radial velocity component vanishes.

it is, suggests that either scenario is dynamically possible depending upon the angular velocity of the solid-body rotation within the core.

Öztekin, Seymour and Varley (2001) have discussed flows involving a vortex core together with a superposed flow sheared normal to the surface. For a class of unsteady flows both updraft and downdraft regions of flow are predicted.

References

Abbott, T. N. G. and Walters, K. 1970 Rheometrical flow systems Part 2. Theory for the orthogonal rheometer, including an exact solution of the Navier–Stokes equations, *J. Fluid Mech.* **40**, 205–213.

Acheson, D. J. 1990 *Elementary Fluid Dynamics*, Oxford: Clarendon Press.

Agrawal, H. L. 1957 A new exact solution of the equations of viscous motion with axial symmetry, *Q. J. Mech. Appl. Math.* **10**, 42–44.

Barenblatt, G. I. 1979 *Similarity, Self-similarity and Intermediate Asymptotics*, New York: Plenum.

Barenblatt, G. I. 1996 *Scaling, Self-similarity and Intermediate Asymptotics*, Cambridge: Cambridge University Press.

Batchelor, G. K. 1951 Note on a class of solutions of the Navier–Stokes equations representing steady rotationally symmetric flow, *Q. J. Mech. Appl. Math.* **4**, 29–41.

Batchelor, G. K. 1954 The skin friction on infinite cylinders moving parallel to their length, *Q. J. Mech. Appl. Math.* **7**, 179–192.

Batchelor, G. K. 1967 *An Introduction to Fluid Dynamics*, Cambridge: Cambridge University Press.

Becker, E. 1976 Laminar film flow on a cylindrical surface, *J. Fluid Mech.* **74**, 297–315.

Bellamy-Knights, P. G. 1970 An unsteady two-cell vortex solution of the Navier–Stokes equations, *J. Fluid Mech.* **41**, 673–687.

Bellamy-Knights, P. G. 1971 Unsteady multicellular viscous vortices, *J. Fluid Mech.* **50**, 1–16.

Beltrami, E. 1889 Considerazione idrodinamiche, *Rend. Ist. Lomb.* (2) **22**, 121–130. Also *Opere* **4**, 300–309.

Berker, R. 1963 Intégration des équations du mouvement d'un fluide visqueux incompressible, *Encyclopedia of Physics*, ed. S. Flügge, vol. VIII/2, Berlin: Springer-Verlag, pp. 1–384.

Berker, R. 1982 An exact solution of the Navier–Stokes equations: the vortex with curvilinear axis, *Int. J. Eng. Sci.* **20**, 217–230.

Berman, A. S. 1953 Laminar flow in channels with porous walls, *J. Appl. Phys.* **24**, 1232–1235.

Berman, A. S. 1958a Laminar flow in an annulus with porous walls, *J. Appl. Phys.* **29**, 71–75.

Berman, A. S. 1958b Effects of porous boundaries on the flow of fluids in systems with various geometries, *Proc. 2nd Int. Conf. on Peaceful Uses of Atomic Energy*, vol. 4, Geneva: UN, pp. 353–358.

Birkhoff, G. 1960 *Hydrodynamics*, 2nd edn, Princeton, NJ: Princeton University Press.

Bluman, G. W. and Kumei, S. 1989 *Symmetries and Differential Equations*, New York: Springer-Verlag.

Blyth, M. G. and Hall, P. 2003 Oscillatory flow near a stagnation point, *SIAM J. Appl. Math.* **63**, 1604–1614.

Blyth, M. G., Hall, P. and Papageorgiu, D. T. 2003 Chaotic flows in pulsating cylindrical tubes: a class of exact Navier–Stokes solutions, *J. Fluid Mech.* **481**, 187–213.

Bödewadt, U. T. 1940 Die Drehströmung über festem Grunde, *Z. Angew. Math. Mech.* **20**, 241–253.

Bodonyi, R. J. 1975 On rotationally symmetric flow above an infinite rotating disk, *J. Fluid Mech.* **67**, 657–666.

Bodonyi, R. J. and Stewartson, K. 1977 The unsteady laminar boundary layer on a rotating disk in a counter-rotating fluid, *J. Fluid Mech.* **79**, 669–688.

Boussinesq, J. 1868 Sur l'influence des frottements dans les mouvements regulières des fluides, *J. Math. Pure Appl.* **13**, 377–438.

Brady, J. F. and Acrivos, A. 1981 Steady flow in a channel or tube with an accelerating surface velocity. An exact solution to the Navier–Stokes equations, *J. Fluid Mech.* **112**, 127–150.

Burde, G. I. 1989 On the motion of fluid near a stretching circular cylinder, *J. Appl. Math. Mech. (PMM.)* **53**, 271–273.

Burde, G. I. 1994 The construction of special explicit solutions of the boundary-layer equations. Steady flows, *Q. J. Mech. Appl. Math.* **47**, 247–260.

Burgers, J. M. 1948 A mathematical model illustrating the theory of turbulence, *Adv. Appl. Mech.* **1**, 171–199.

Carslaw, H. S. and Jaeger, J. C. 1947 *Conduction of Heat in Solids*, Oxford: Oxford University Press.

Cheng, E. H. W., Özişik, M. N. and Williams III, J. C. 1971 Nonsteady three-dimensional stagnation-point flow, *J. Appl. Mech.* **38**, 282–287.

Cochran, W. G. 1934 The flow due to a rotating disc, *Proc. Cambridge Philos. Soc.* **30**, 365–375.

Couette, M. 1890 Études sur le frottement des liquides, *Ann. Chim. Phys.* (6) **21**, 433–510.

Coward, A. V. and Hall, P. 1996 The stability of two-phase flow over a swept wing, *J. Fluid Mech.* **329**, 247–273.

Cox, S. M. 1991a Analysis of steady flow in a channel with one porous wall, or with accelerating walls, *SIAM J. Appl. Math.* **51**, 429–438.

Cox, S. M. 1991b Two-dimensional flow of a viscous fluid in a channel with porous walls, *J. Fluid Mech.* **227**, 1–33.

Cox, S. M. 2002 Nonaxisymmetric flow between an air table and a floating disk, *Phys. Fluids* **14**, 1540–1543.

Craik, A. D. D. and Criminale, W. O. 1986 Evolution of wavelike disturbances in shear flows: a class of exact solutions of the Navier–Stokes equations, *Proc. R. Soc. London, Ser. A* **406**, 13–26.

Crane, L. J. 1970 Flow past a stretching plate, *Z. Angew. Math. Phys.* **21**, 645–647.

Cunning, G. M., Davis, A. M. J. and Weidman, P. D. 1998 Radial stagnation flow on a rotating circular cylinder with uniform transpiration, *J. Eng. Math.* **33**, 113–128.

Danberg, J. E. and Fansler, K. S. 1976 A nonsimilar moving-wall boundary-layer problem, *Q. Appl. Math.* **33**, 305–309.

Dauenhauer, E. C. and Majdalani, J. 2003 Exact self-similarity solution of the Navier–Stokes equations for a porous channel with orthogonally moving walls, *Phys. Fluids* **15**, 1485–1495.

Davey, A. 1961 Boundary-layer flow at a saddle point of attachment, *J. Fluid Mech.* **10**, 593–610.

Davey, A. 1963 Rotational flow near a forward stagnation point, *Q. J. Mech. Appl. Math.* **16**, 33–59.

Davey, A. and Schofield, D. 1967 Three-dimensional flow near a two-dimensional stagnation point, *J. Fluid Mech.* **28**, 149–151.

Dean, W. R. 1934 Note on the divergent flow of fluid, *Philos. Mag.* (7) **18**, 759–777.

Debler, W. R. and Montgomery, R. D. 1971 Flow over an oscillating plate with suction or with an intermediate film: two exact solutions of the Navier–Stokes equations, *J. Appl. Mech.* **38**, 262–265.

Devi, C. D. S., Takhar, H. S. and Nath, G. 1986 Unsteady, three-dimensional, boundary-layer flow due to a stretching surface, *Int. J. Heat Mass Transfer* **29**, 1996–1999.

Dijkstra, D. 1980 On the relation between adjacent inviscid cell type solutions to the rotating-disk equations, *J. Eng. Math.* **14**, 133–154.

Dijkstra, D. and Zandbergen, P. J. 1978 Some further investigations on nonunique solutions of the Navier–Stokes equations for the Kármán swirling flow, *Arch. Mech. Stosow.* **30**, 411–419.

van Dommelen, L. L. and Shen, S. F. 1980 The spontaneous generation of the singularity in a separating laminar boundary layer, *J. Comput. Phys.* **38**, 125–140.

Donaldson, C. du P. and Sullivan, R. D. 1960 Examination of the solutions of the Navier–Stokes equations for a class of three-dimensional vortices. Part I. Velocity distributions for steady motion, *Aero. Res. Assoc. Princeton Rep.* AFOSR TN 60–1227.

Dorrepaal, M. J. 1986 An exact solution of the Navier–Stokes equation which describes non-orthogonal stagnation-point flow in two dimensions, *J. Fluid Mech.* **163**, 141–147.

Drazin, P. G. 2002 *Introduction to Hydrodynamic Stability*, Cambridge: Cambridge University Press.

Drazin, P. G. and Reid, W. H. 1981 *Hydrodynamic Stability*, Cambridge: Cambridge University Press.

Dryden, H. L., Murnaghan, F. D. and Bateman, H. 1932 Hydrodynamics, *Bulletin of the National Research Council, Washington*, no. 84. (Reprinted in 1956 by Dover Publications, New York.)

Durlofsky, L. and Brady, J. F. 1984 The spatial stability of a class of similarity solutions, *Phys. Fluids* **27**, 1068–1076.

Ekman, V. W. 1905 On the influence of the Earth's rotation on ocean currents, *Arkiv. Mat. Astron. Fys.* **2**, 1–52.

Evans, D. J. 1969 The rotationally symmetric flow of a viscous fluid in the presence of an infinite rotating disk with uniform suction, *Q. J. Mech. Appl. Math.* **22**, 467–485.

Fraenkel, L. E. 1962 Laminar flow in symmetrical channels with slightly curved walls I. On the Jeffery-Hamel solutions for flow between plane walls, *Proc. R. Soc. London, Ser. A* **267**, 119–138.

Fraenkel, L. E. 1963 Laminar flow in symmetrical channels with slightly curved walls II. An asymptotic series for the stream function, *Proc. R. Soc. London, Ser. A* **272**, 406–428.

Fushschich, W. I., Shtelen, W. M. and Serov, N. I. 1993 *Symmetry Analysis and Exact Solutions of Equations of Mathematical Physics* Dordrecht: Kluwer.

Gibbon, J. D., Fokas, A. S. and Doering, C. R. 1999 Dynamically stretched vortices as solutions of the 3D Navier–Stokes equations, *Physica D* **132**, 497–510.

Glauert, M. B. 1956 The laminar boundary layer on oscillating plates and cylinders, *J. Fluid Mech.* **1**, 97–110.

Goldshtik, M. A. 1960 A paradoxical solution of the Navier–Stokes equations, *Prik. Mat. Mekh.* **24**, 610–621 (in Russian).

Goldshtik, M. A. 1990 Viscous flow paradoxes, *Annu. Rev. Fluid Mech.* **22**, 441–472.

Goldshtik, M. A. and Shtern, V. N. 1989 Loss of symmetry in viscous flow from a linear source, *Fluid Dyn.* **24**, 151–199.

Goldshtik, M. A. and Shtern, V. N. 1990 Collapse in conical viscous flows, *J. Fluid Mech.* **218**, 483–508.

Goldstein, S. (ed.) 1938 *Modern Developments in Fluid Dynamics*, Oxford: Oxford University Press.

Gorla, R. S. R. 1978 Nonsimilar axisymmetric stagnation flow on a moving cylinder, *Int. J. Eng. Sci.* **16**, 397–400.

Gorla, R. S. R. 1979 Unsteady viscous flow in the vicinity of an axisymmetric stagnation point on a circular cylinder, *Int. J. Eng. Sci.* **17**, 87–93.

Görtler, H. 1944 Verdrängungswirkung der laminaren Grenzschichten und Druck widerstand, *Ing. Arch.* **14**, 286–305.

Grauel, A. and Steeb, W. H. 1985 Similarity solutions of the Euler equation and the Navier–Stokes equation in two space dimensions, *Int. J. Theor. Phys.* **24**, 255–265.

Gromeka, I. 1881 Some cases of incompressible fluid motion, *Collected. Works*, Kazan (in Russian).

Grosch, C. E. and Salwen, H. 1982 Oscillating stagnation point flow, *Proc. R. Soc. London, Ser. A* **384**, 175–190.

Grundy, R. E. and McLaughlin, R. 1999 Three-dimensional blow-up solutions of the Navier–Stokes equations, *IMA J. Appl. Math.* **63**, 287–306.

Hagen, G. 1839 Über die Bewegung des Wassers in engen cylindrischen Rohren, *Pogg. Ann.* **46**, 423–442.

Hall, P. and Papageorgiu, D. T. 1999 The onset of chaos in a class of Navier–Stokes solutions, *J. Fluid Mech.* **393**, 59–87.

Hall, P., Balakumar, P. and Papageorgiu, D. 1992 On a class of unsteady three-dimensional Navier–Stokes solutions relevant to rotating disc flows: threshold amplitudes and finite-time singularities, *J. Fluid Mech.* **238**, 297–323.

Hamel, G. 1916 Spiralförmige Bewegungen zäher Flüssigkeiten, *Jber. dtsch. Math.-Ver.* **25**, 34–60. Translated as NACA Tech. Memo. no. 1342 (1953).

Hamza, E. A. and MacDonald, D. A. 1984 A similar flow between two rotating disks, *Q. Appl. Math.* **41**, 495–511.

Hannah, D. M. 1947 Forced flow against a rotating disc, *Rep. Mem. Aero. Res. Council, London* No. 2772.

Harrison, W. J. 1919 The pressure in a viscous liquid moving through a channel with diverging boundaries, *Proc. Cambridge Philos. Soc.* **19**, 307–312.

Hasimoto, H. 1951 Note on Rayleigh's problem for a bent flat plate, *J. Phys. Soc. Jpn.* **6**, 400–401.

Hasimoto, H. 1956 Note on Rayleigh's problem for a circular cylinder with uniform suction and related unsteady flow problem, *J. Phys. Soc. Jpn.* **11**, 611–612.

Hasimoto, H. 1957 Boundary layer growth on a flat plate with suction or injection, *J. Phys. Soc. Jpn.* **12**, 68–72.

Hatton, L. 1975 Stagnation point flow in a vortex core, *Tellus* **27**, 269–280.

Hersh, A. S. 1970 Unsteady viscous rotational stagnation-point flow, *Z. Angew. Math. Phys.* **21**, 1093–1096.

Hewitt, R. E., Duck, P. W. and Foster, M. R. 1999 Steady boundary-layer solutions for a swirling stratified fluid in a rotating cone, *J. Fluid Mech.* **384**, 339–374.

Hewitt, R. E., Duck, P. W. and Stow, S. R. 2002 Continua of states in boundary-layer flows, *J. Fluid Mech.* **468**, 121–152.

Hiemenz, K. 1911 Die Grenschicht an einem in den gleichförmigen Flüssigkeitsstrom eingetauchten geraden Kreiszlinder, *Dinglers Polytech. J.* **326**, 321–324, 344–348, 357–362, 372–376, 391–393, 407–410.

Hill, M. J. M. 1894 On a spherical vortex, *Philos. Trans. R. Soc. London, Ser. A* **185**, 213–245.

Hinch, E. J. and Lemaître, J. 1994 The effect of viscosity on the height of disks floating above an air table, *J. Fluid Mech.* **273**, 313–322.

Hinze, J. and van der Hegge Zijnen, B. 1949 Transfer of heat and matter in the turbulent mixing zone of an axially symmetrical jet, *Appl. Sci. Res.* **1**, 435–461.

Hocking, L. M. 1963 An example of boundary layer formation, *AIAA J.* **1**, 1222–1223.

Holodniok, M., Kubicek, M. and Hlavacek, V. 1981 Computation of the flow between two rotating coaxial disks: multiplicity of steady-state solutions, *J. Fluid Mech.* **108**, 227–240.

Homann, F. 1936 Der Einfluss grosser Zähigkeit bei der Strömung um den Zylinder und um die Kugel, *Z. Angew. Math. Mech.* **16**, 153–164.

Hooper, A. P., Duffy, B. R. and Moffatt, H. K. 1982 Flow of fluid of non-uniform viscosity in converging and diverging channels, *J. Fluid Mech.* **117**, 283–304.

Howarth, L. 1934 On calculation of the steady flow in the boundary layer near the surface of a cylinder in a stream, *Rep. ARC*, 1632.

Howarth, L. 1950 Rayleigh's problem for a semi-infinite plate, *Proc. Cambridge Philos. Soc.* **46**, 127–140.

Howarth, L. 1951 The boundary layer in three-dimensional flow. Part II: The flow near a stagnation point, *Philos. Mag.* (7) **42**, 1433–1440.

Hui, W. H. 1987 Exact solutions of the unsteady two-dimensional Navier–Stokes equations, *Z. Angew. Math. Phys.* **38**, 689–702.

Irmay, S. and Zuzovsky, M. 1970 Exact solutions of the Navier–Stokes equations in two-way flows, *Isr. J. Technol.* **8**, 307–315.

Jeffery, G. B. 1915 The two-dimensional steady motion of a viscous fluid, *Philos. Mag.* (6) **29**, 455–465.

Kambe, T. 1983 A class of exact solutions of two-dimensional viscous flow, *J. Phys. Soc. Jpn.* **52**, 834–841.

Kambe, T. 1986 A class of exact solutions of the Navier–Stokes equation, *Fluid Dyn. Res.* **1**, 21–31.

von Kármán, T. 1921 Über laminare und turbulente Reibung, *Z. Angew. Math. Mech.* **1**, 233–252.

von Kármán, T. 1924 Über die Oberflächenreibung von Flüssigkeiten, *Vorträge aus dem Gebiete der Hydro-und Aerodynamik*, ed. von Kármán, T. and Levi-civita, T., Berlin: Springer, pp. 150–151.

Keller, H. B. and Szeto, R. K. H. 1980 Calculation of flows between rotating disks. In *Computing Methods in Applied Sciences and Engineering*, ed. Glowinski, R. and Lions, J. L., Amsterdam: North Holland, pp. 51–61.

Kelly, R. E. 1965 The flow of a viscous fluid past a wall of infinite extent with time-dependent suction, *Q. J. Mech. Appl. Math.* **18**, 287–298.

Kelvin, Lord 1887 Stability of fluid motion. Rectilineal motion of viscous fluid between two parallel planes, *Philos. Mag.* **24**, 188–196.

Kerr, O. S. and Dold, J. W. 1994 Periodic steady vortices in a stagnation-point flow, *J. Fluid Mech.* **276**, 307–325.

Kovásznay, L. I. G. 1948 Laminar flow behind a two-dimensional grid, *Proc. Cambridge Philos. Soc.* **44**, 58–62.

Kreiss, H. O. and Parter, S. V. 1983 On the swirling flow between rotating coaxial disks: existence and nonuniqueness, *Commun. Pure Appl. Math.* **36**, 55–84.

Lagerstrom, P. A. 1996 *Laminar Flow Theory*, Princeton, NJ: Princeton University Press. (Originally published in *Theory of Laminar Flows*, ed. F. K. Moore, Princeton, NJ: Princeton University Press, 1964.)

Lagerstrom, P. A. and Cole, J. D. 1955 Examples illustrating expansion procedures for the Navier–Stokes equations, *J. Rat. Mech. Anal.* **4**, 817–882.

Lagnado, R. R. and Leal, L. G. 1990 Visualization of three-dimensional flow in a four-roll mill, *Exps. Fluids* **9**, 25–32.

Lambossy, P. 1952 Oscillations forcées d'un liquide incompressible et visqueux dans un tube rigide et horizontal. Calcul de la force de frottement, *Helv. Phys. Acta* **25**, 371–386.

Lance, G. N. 1956 Motion of a viscous fluid in a tube which is subjected to a series of pulses, *Q. Appl. Math.* **14**, 312–315.

Lance, G. N. and Rogers, M. H. 1962 The axially symmetric flow of a viscous fluid between two infinite rotating disks, *Proc. R. Soc. London, ser. A.* **266**, 109–121.

Landau, L. D. 1944 A new exact solution of the Navier–Stokes equations, *Dokl. Akad. Nauk SSSR* **43**, 286–288. See also Landau, L. M. and Lifshitz, E. M. 1959 *Fluid Mechanics*, London: Pergamon, pp. 86–88.

Langlois, W. E. 1985 Buoyancy-driven flows in crystal-growth melts, *Annu. Rev. Fluid Mech.* **17**, 191–215.

Lawrence, C. J., Kuang, Y. and Weinbaum, S. 1985 The inertial draining of a thin fluid layer between parallel plates with a constant normal force. Part 2. Boundary layer and exact numerical solutions, *J. Fluid Mech.* **156**, 479–494.

Leray, J. 1934 Sur le mouvement d'un liquide visqueux emplissante l'espace, *Acta Math.* **63**, 193–248.

Levine, H. 1957 Skin friction on a strip of finite width moving parallel to its length, *J. Fluid Mech.* **3**, 145–158.

Libby, P. A. 1967 Heat and mass transfer at a general three-dimensional stagnation point, *AIAA J.* **5**, 507–517.

Libby, P. A. 1974 Wall shear at a three-dimensional stagnation point with a moving wall, *AIAA J.* **12**, 408–409.

Libby, P. A. 1976 Laminar flow at a three-dimensional stagnation point with large rates of injection, *AIAA J.* **14**, 1273–1279.

Lin, S. P. and Tobak, M. 1986 Reversed flow above a plate with suction, *AIAA J.* **24**, 334–335.

Long, R. R. 1958 Vortex motion in a viscous fluid, *J. Meteorol.* **15**, 108–112.

Long, R. R. 1961a *Mechanics of Solids and Fluids*, Englewood Cliffs, NJ: Prentice-Hall.

Long, R. R. 1961b A vortex in an infinite viscous fluid, *J. Fluid Mech.* **11**, 611–624.

Ludlow, D. K., Clarkson, P. A. and Bassom, A. P. 1998 Nonclassical symmetry reductions of the three-dimensional incompressible Navier–Stokes equations, *J. Phys. A: Math. Gen.* **31**, 7965–7980.

Ludlow, D. K., Clarkson, P. A., and Bassom, A.P. 1999 Similarity reductions and exact solutions for the two-dimensional incompressible Navier–Stokes equations, *Stud. Appl. Math.* **103**, 183–240.

McLeod, J. B. 1971 The existence of axially symmetric flow above a rotating disk, *Proc. R. Soc. London, Ser. A* **324**, 391–414.

McLeod, J. B. and Parter, S. V. 1974 On the flow between two counter-rotating infinite plane disks, *Arch. Rat. Mech. Anal.* **54**, 301–327.

McLeod, J. B. and Rajagopal, K. R. 1987 On uniqueness of flow of a Navier–Stokes fluid due to a stretching boundary, *Arch. Rat. Mech. Anal.* **42**, 385–393.

Mallock, A. 1896 Experiments on fluid viscosity, *Philos. Trans. R. Soc. London, Ser. A* **187**, 41–56.

Marques, F., Sanchez, J. and Weidman, P. D. 1998 Generalized Couette–Poiseuille flow with boundary mass transfer, *J. Fluid Mech.* **374**, 221–249.

Marris, A. W. and Aswani, M. G. 1977 On the general impossibility of controllable axisymmetric Navier–Stokes motions, *Arch. Rat. Mech. Anal.* **63**, 107–153.

Mellor, G. L., Chapple, P. J. and Stokes, V. K. 1968 On the flow between a rotating and a stationary disk, *J. Fluid Mech.* **31**, 95–112.

Merchant, G. J. and Davis, S. H. 1989 Modulated stagnation-point flow and steady streaming, *J. Fluid Mech.* **198**, 543–555.

Meshalkin, L. D. and Sinai, I. G. 1961 Investigation of the stability of a stationary solution of a system of equations for the plane movement of an incompressible fluid, *J. Appl. Math. Mech.* **25**, 1700–1705.

Millsaps, K. and Pohlhausen, K. 1953 Thermal distributions in Jeffery–Hamel flows between nonparallel plane walls, *J. Aerosp. Sci.* **20**, 187–196.

Moffatt, H. K. 2000 The interaction of skewed vortex pairs: a model for blow-up of the Navier–Stokes equations, *J. Fluid Mech.* **409**, 51–68.

Moffatt, H. K. and Duffy, B. R. 1980 Local similarity solutions and their limitations, *J. Fluid Mech.* **96**, 299–313 (and Corrigendum **99**, 1980, 860).

Moore, R. L. 1991 Exact non-linear forced periodic solutions of the Navier–Stokes equation, *Physica D* **52**, 179–190.

Morgan, A. J. A. 1956 On a class of laminar viscous flows within one or two bounding cones, *Aeronaut. Q.* **7**, 225–239.

Morton, B. R. 1966 Geophysical vortices, *Prog. Aeronaut. Sci.* **7**, 145–193.

Navier, C.-L.-M.-H. 1827 Mémoire sur les lois du mouvement des fluides (1822), *Mem. Acad. Sci. Inst. France* (2) **6**, 389–440.

Newton, I. 1687 *Philosophiae Naturalis Principia Mathematica*, London: Royal Society.

Noether, F. 1931 *Handbuch der Physikalischen und Technischen Mechanik*, vol. 5, Leipzig: J. A. Barch, pp. 733–736.

O'Brien, V. 1961 Steady spheroidal vortices – more exact solutions to the Navier–Stokes equation, *Q. Appl. Math.* **19**, 163–168.

Ockendon, H. 1972 An asymptotic solution for steady flow above an infinite rotating disk with suction, *Q. J. Mech. Appl. Math.* **25**, 291–301.

Olver, P. 1992 *Applications of Lie Groups to Differential Equations*, 2nd edn, New York: Springer-Verlag.

Oseen, C. W. 1911 Über Wirbelbewegung in einer reibenden Flüssigkeit, *Ark. Mat. Astron. Fys.* **7**, 14–26.

Öztekin, A., Seymour, B. R. and Varley, E. 2001 Pump flow solutions of the Navier–Stokes equations, *Stud. Appl. Math.* **107**, 1–41.

Paull, R. and Pillow, A. F. 1985a Conically similar viscous flows. Part 2. One-parameter swirl-free flows, *J. Fluid Mech.* **155**, 343–358.

Paull, R. and Pillow, A. F. 1985b Conically similar viscous flows. Part 3. Characterization of axial causes in swirling flow and the one-parameter flow generated by a uniform half-line source of kinematic swirl angular momentum, *J. Fluid Mech.* **155**, 359–379.

Pearson, C. E. 1965 Numerical solutions for the time-dependent viscous flow between two rotating coaxial disks, *J. Fluid Mech.* **21**, 623–633.

Pesch, H. J. and Rentrop, P. 1978 Numerical solution of the flow between two counter-rotating infinite plane disks by multiple shooting, *Z. Angew. Math. Mech.* **58**, 23–28.

Pillow, A. F. and Paull, R. 1985 Conically similar viscous flows. Part I. Basic conservation principles and characterization of axial causes in swirl-free flow, *J. Fluid Mech.* **155**, 327–341.

Poiseuille, J. L. M. 1840 Recherches expérimentales sur le mouvement des liquides dans les tubes de très petits diametres, *C. R. Acad. Sci., Paris* **11**, 961–967, 1041–1048, **12**, 112–115.

Poisson, S.-D. 1831 Mémoire sur les équations générales de l'équilibre et du mouvement des corps solides élastique et des fluides (1829), *J. Ec. Polytech.* **13**, cahier 20, 1–174.

Prager, S. 1964 Spiral flow in a stationary porous pipe, *Phys. Fluids* **7**, 907–908.

Preston, J. H. 1950 The steady circulatory flow about a circular cylinder with uniformly distributed suction at the surface, *Aeronaut Q.* **1**, 319–338.

Proudman, J. 1914 Notes on the motion of viscous liquids in channels, *Philos. Mag.* (5) **28**, 30–36, 760.

Proudman, I. and Johnson, K. 1962 Boundary-layer growth near a rear stagnation point, *J. Fluid Mech.* **12**, 161–168.

Putkaradze, V. 2003 Radial flow of two immiscible fluids: analytical solutions and bifurcations, *J. Fluid Mech.* **477**, 1–18.

Putkaradze, V. and Dimon, P. 2000 Nonuniform two-dimensional flow from a point source, *Phys. Fluids* **12**, 66–70.

Raithby, G. D. 1971 Laminar heat transfer in the thermal entrance region of circular tubes and two-dimensional rectangular ducts with wall suction and injection, *Int. J. Heat Mass Transfer* **14**, 223–243.

Rajappa, N. R. 1979 Nonsteady plane stagnation point flow with hard blowing, *Z. Angew. Math. Mech.* **59**, 471–473.

Rao, A. R. and Kasiviswanathan, S. R. 1987 On exact solutions of the unsteady Navier–Stokes equations – the vortex with instantaneous curvilinear axis, *Int. J. Eng. Sci.* **25**, 337–349.

Rayleigh, Lord 1911 On the motion of solid bodies through viscous liquids, *Philos. Mag.* **21**, 697–711.

Riabouchinsky, D. 1924 Quelques considérations sur les mouvements plans rotationnels d'un liquide, *C. R. Acad. Sci., Paris* **179**, 1133–1136.

Riley, N. 1987 Boundary-layer models of magnetic Czochralski crystal growth, *J. Cryst. Growth* **85**, 417–421.

Riley, N. and Vasantha, R. 1989 An unsteady stagnation-point flow, *Q. J. Mech. Appl. Math.* **42**, 511–521.

Robins, A. J. and Howarth, J. A. 1972 Boundary-layer development at a two-dimensional rear stagnation point, *J. Fluid Mech.* **56**, 161–171.

Robinson, A. C. and Saffman, P. G. 1984 Stability and structure of stretched vortices, *Stud. Appl. Math.* **70**, 163–181.

Robinson, W. A. 1976 The existence of multiple solutions for the laminar flow in a uniformly porous channel with suction at both walls, *J. Eng. Math.* **10**, 23–40.

Rogers, M. H. and Lance, G. N. 1960 The rotationally symmetric flow of a viscous fluid in the presence of an infinite rotating disk, *J. Fluid Mech.* **7**, 617–631.

Rosenhead, L. 1940 The steady two-dimensional radial flow of viscous fluid between two inclined plane walls, *Proc. R. Soc. London, Ser. A* **175**, 436–467.

Rott, N. 1956 Unsteady viscous flow in the vicinity of a stagnation point, *Q. Appl. Math.* **13**, 444–451.

Rott, N. 1958 On the viscous core of a line vortex, *Z. Angew. Math. Phys.* **9**, 543–553.

Rott, N. and Lewellen, W. S. 1967 Boundary layers due to the combined effects of rotation and translation, *Phys. Fluids* **10**, 1867–1873.

Rowell, H. S. and Finlayson, D. 1928 Screw viscosity pumps, *Engineering, London* **126**, 249–250, 385–387. (See also Berker 1963, p. 71.)

Saint-Venant, B. 1843 Mémoire sur la dynamique des fluides, *C. R. Acad. Sci., Paris* **17**, 1240–1242.

Sanyal, L. 1956 The flow of a viscous liquid in a circular tube under pressure-gradients varying exponentially with time, *Indian J. Phys.* **30**, 57–61.

Schlichting, H. 1979 *Boundary-Layer Theory*, 7th edn, London: Pergamon Press.

Schofield, D. and Davey, A. 1967 Dual solutions of the boundary-layer equations at a point of attachment, *J. Fluid Mech.* **30**, 809–811.

Sedov, L. 1959 *Similarity and Dimensional Methods in Mechanics*, 6th edn, New York: Academic Press.

Serrin, J. 1972 The swirling vortex, *Philos. Trans. R. Soc. London, Ser. A.* **271**, 325–360.

Sexl, T. 1930 Uber den von E. G. Richardson entdeckten 'Annulareffekt', *Z. Phys.* **61**, 349–362.

Shih, K.-G. 1987 On the existence of solutions of an equation arising in the theory of laminar flow in a uniformly porous channel with injection, *SIAM J. Appl. Math.* **47**, 526–533.

Shrestha, G. M. and Terrill, R. M. 1968 Laminar flow with large injection through parallel and uniformly porous walls of different permeability, *Q. J. Mech. Appl. Math.* **21**, 413–432.

Skalak, F. M. and Wang, C. Y. 1977 Pulsatile flow in a tube with wall injection and suction, *Appl. Sci. Res.* **33**, 269–307.

Skalak, F. M. and Wang, C. Y. 1978 On the nonunique solutions of laminar flow through a porous tube or channel, *SIAM J. Appl. Math.* **34**, 535–544.

Skalak, F. M. and Wang, C. Y. 1979 On the unsteady squeezing of a viscous fluid from a tube, *J. Aust. Math. Soc. B* **21**, 65–74.

Slezkin, N. A. 1934 On an exact solution of the equations of viscous flow, *Uch. Zap. Mosk. Gos. Univ., Sci. Rec. Moscow State Univ.* **2**, 89–90 (in Russian). See also *Prik. Mat. Mekh.* **18**, 1954, 764 (in Russian).

Smith, S. H. 1987 Eccentric rotating flows: exact unsteady solutions of the Navier–Stokes equations, *Z. Angew. Math. Phys.* **38**, 573–579.

Smith, S. H. 1997 Time-dependent Poiseuille flow, *SIAM Rev.* **39**, 511–513.

Sowerby. L. 1951 The unsteady flow of a viscous incompressible fluid inside an infinite channel, *Philos. Mag.* **42**, 176–187.

Sowerby, L. and Cooke, J. C. 1953 The flow of fluid along corners and edges, *Q. J. Mech. Appl. Math.* **6**, 50–70.

Sozou, C. 1992 On solutions relating to conical vortices over a plane wall, *J. Fluid Mech.* **244**, 633–644.

Spencer, A. J. M. 1980 *Continuum Mechanics*, London: Longman.

Squire, H. B. 1951 The round laminar jet, *Q. J. Mech. Appl. Math.* **4**, 321–329.

Squire, H. B. 1952 Some viscous flow problems I. Jet emerging from a hole in a plane wall, *Philos. Mag.* **43**, 942–945.

Squire, H. B. 1955 Radial jets, *50 Jahre Grenzschichtforschung*, ed. Görtler, H. and Tollmien, W., Braunschweig, Vieweg, pp. 47–54.

Stein, C. F. 2000 On the existence of conically self-similar free-vortex solutions to the Navier–Stokes equations, *IMAJ Appl. Math.* **64**, 283–300.

Stein, C. F. 2001 Quantities which define conically self-similar free-vortex solutions to the Navier–Stokes equations uniquely, *J. Fluid Mech.* **438**, 159–181.

Stewartson, K. 1953 On the flow between two rotating coaxial disks, *Proc. Cambridge Philos. Soc.* **49**, 333–341.

Stokes, G. G. 1842 On the steady motion of incompressible fluids, *Trans. Cambridge Philos. Soc.* **7**, 439–453. Also *Math. Phys. Papers* **1**, 1880, 1–16.

Stokes, G. G. 1845 On the theories of the internal friction of fluids in motion, and of the equilibrium and motion of elastic solids, *Trans. Cambridge Philos. Soc.* **8**, 287–305. Also *Math. Phys. Papers* **1**, 1880, 75–129.

Stokes, G. G. 1846 Report on recent researches in hydrodynamics, *Rep. Br. Assoc.*, 1–20. Also *Math. Phys. Papers* **1**, 1880, 156–187.

Stokes, G. G. 1851 On the effect of the internal friction of fluids on the motion of pendulums, *Trans. Cambridge Philos. Soc.* **9**, 8–106. Also *Math. Phys. Papers* **3**, 1880, 1–141.

Stuart, J. T. 1954 On the effects of uniform suction on the steady flow due to a rotating disk, *Q. J. Mech. Appl. Math.* **7**, 446–457.

Stuart, J. T. 1955 A solution of the Navier–Stokes and energy equations illustrating the response of skin friction and temperature of an infinite plate thermometer to fluctuations in the stream velocity, *Proc. R. Soc. London, Ser. A* **231**, 116–130.

Stuart, J. T. 1959 The viscous flow near a stagnation point when the external flow has uniform vorticity, *J. Aerosp. Sci.* **26**, 124–125.

Stuart, J. T. 1966a Double boundary layers in oscillating viscous flows, *J. Fluid Mech.* **24**, 673–687.

Stuart, J. T. 1966b A simple corner flow with suction, *Q. J. Mech. Appl. Math.* **19**, 217–220.

Sullivan, R. D. 1959 A two-cell vortex solution of the Navier–Stokes equations, *J. Aerosp. Sci.* **26**, 767–768.

Szymański, P. 1932 Quelques solutions exactes des équations de l'hydrodynamique du fluide visqueux dans le cas d'un tube cylindrique, *J. Math. Pures Appl.* (9) **11**, 67–107.

Tam, K. K. 1969 A note on the asymptotic solution of the flow between two oppositely rotating infinite plane disks, *SIAM J. Appl. Math.* **17**, 1305–1310.

Tamada, K. 1979 Two-dimensional stagnation-point flow impinging obliquely on a plane wall, *J. Phys. Soc. Jpn.* **46**, 310–311.

Taylor, C. L., Banks, W. H. H., Zaturska, M. B. and Drazin, P. G. 1991 Three-dimensional flow in a porous channel, *Q. J. Mech. Appl. Math.* **44**, 105–133.

Taylor, G. I. 1923 On the decay of vortices in a viscous fluid, *Philos. Mag.* **46**, 671–674.

Terrill, R. M. 1964 Laminar flow in a uniformly porous channel, *Aeronaut. Q.* **15**, 299–310.

Terrill, R. M. 1965 Laminar flow in a uniformly porous channel with large injection, *Aeronaut. Q.* **16**, 323–332.

Terrill, R. M. 1966 Flow through a porous annulus, *Appl. Sci. Res.* **17**, 204–222.

Terrill, R. M. 1973 On some exponentially small terms arising in the flow through a porous pipe, *Q. J. Mech. Appl. Math.* **26**, 347–354.

Terrill, R. M. 1982 An exact solution for flow in a porous pipe, *Z. Angew. Math. Phys.* **33**, 547–552.

Terrill, R. M. and Shrestha, G. M. 1965 Laminar flow through parallel and uniformly porous walls of different permeabilities, *Z. Angew. Math. Phys.* **16**, 470–482.

Terrill, R. M. and Thomas, P. W. 1969 On laminar flow through a uniformly porous pipe, *Appl. Sci. Res.* **21**, 37–67.

Terrill, R. M. and Thomas, P. W. 1973 Spiral flow in a porous pipe, *Phys. Fluids* **16**, 356–359.

Thorpe, J. F. 1967 Further investigation of squeezing flow between parallel plates, *Dev. Theor. Appl. Mech.* **3**, 635–648.

Tifford, A. N. and Chu, S. T. 1952 Uniform stream past rotating disk, *J. Aerosp. Sci.* **19**, 284–285.

Tilley, B. S. and Weidman, P. D. 1998 Oblique two-fluid stagnation-point flow, *Eur. J. Mech. B, Fluids* **17**, 205–217.

Tollmien, W. 1921 Grenzschichttheorie, *Handbuch der Experimentalphysik*, vol. 4, Teil I, Leipzig: Akademische Verlagsgesellschaft, pp. 241–287.

Tranter, C. J. 1951 Heat conduction in the region bounded internally by an elliptic cylinder, and an analogous problem in atmospheric diffusion, *Q. J. Mech. Appl. Math.* **4**, 461–465.

Trkal, V. 1919 A remark on the hydrodynamics of viscous fluids, *Cas. Pst. Mat. Fys.* **48**, 302–311 (in Czech).

Truesdell, C. A. 1954 *The Kinematics of Vorticity*, Bloomington, IN: Indiana University Press.

Tsien, H. S. 1943 Symmetrical Joukowsky airfoils in shear flow, *Q. Appl. Math.* **1**, 130–148.

Uchida, S. 1956 The pulsating viscous flow superposed on the steady laminar motion of incompressible fluid in a circular pipe, *Z. Angew. Math. Mech.* **7**, 403–421.

Uchida, S. and Aoki, H. 1977 Unsteady flows in a semi-infinite contracting or expanding pipe, *J. Fluid Mech.* **82**, 371–387.

Wang, C. Y. 1966 On a class of exact solutions of the Navier–Stokes equations, *J. Appl. Mech.* **33**, 696–698.

Wang, C. Y. 1971 Pulsatile flow in a porous channel, *J. Appl. Mech.* **38**, 553–555.

Wang, C. Y. 1973 Axisymmetric stagnation flow towards a moving plate, *Am. Inst. Chem. Eng. J.* **19**, 1080–1081.

Wang, C. Y. 1974 Axisymmetric stagnation flow on a cylinder, *Q. Appl. Math.* **32**, 207–213.

Wang, C. Y. 1976 The squeezing of a fluid between two plates, *J. Appl. Mech.* **43**, 579–582.

Wang, C. Y. 1984 The three-dimensional flow due to a stretching flat surface, *Phys. Fluids* **27**, 1915–1917.

Wang, C. Y. 1985a Stagnation flow on the surface of a quiescent fluid – an exact solution of the Navier–Stokes equations, *Q. Appl. Math.* **43**, 215–223.

Wang, C. Y. 1985b The unsteady oblique stagnation point flow, *Phys. Fluids* **28**, 2046–2049.

Wang, C. Y. 1987 Impinging stagnation flows, *Phys. Fluids* **30**, 915–917.

Wang, C. Y. 1988 Fluid flow due to a stretching cylinder, *Phys. Fluids* **31**, 466–468.

Wang, C. Y. 1989a Exact solutions of the unsteady Navier–Stokes equations, *Appl. Mech. Rev.* **42**, S269–S282.

Wang, C. Y. 1989b Shear flow over a rotating plate, *Appl. Sci. Res.* **46**, 89–96.

Wang, C. Y. 1990a Exact solution of the Navier–Stokes equations – the generalized Beltrami flows, review and extension, *Acta Mech.* **81**, 69–74.

Wang, C. Y. 1990b Shear flow over convection cells – an exact solution of the Navier–Stokes equations, *Z. Angew. Math. Mech.* **70**, 351–352.

Wang, C. Y. 1991 Exact solutions of the steady-state Navier–Stokes equations, *Annu. Rev. Fluid Mech.* **23**, 159–177.

Wang, C. Y. and Watson, L. T. 1979 Squeezing of a viscous fluid between elliptic plates, *Appl. Sci. Res.* **35**, 195–207.

Wang, C. Y., Watson, L. T. and Alexander, K. A. 1991 Spinning of a liquid film from an accelerating disk, *IMA J. Appl. Math.* **46**, 201–210.

Waters, S. L. 2001 Solute uptake through the walls of a pulsating channel, *J. Fluid Mech.* **433**, 193–208.

Watson, E. J. 1955 Boundary-layer growth, *Proc. R. Soc. London, Ser. A.* **231**, 104–116.

Watson, J. 1958 A solution of the Navier–Stokes equations illustrating the response of a laminar boundary layer to a given change in the external stream velocity, *Q. J. Mech. Appl. Math.* **11**, 302–325.

Watson, J. 1959 The two-dimensional laminar flow near the stagnation point of a cylinder which has an arbitrary transverse motion, *Q. J. Mech. Appl. Math.* **12**, 175–190.

Watson, J. 1966 On the existence of solutions for a class of rotating disc flows and the convergence of a successive approximation scheme, *J. Inst. Math. Appl.* **1**, 348–371.

Watson, L. T. and Wang, C. Y. 1979 Deceleration of a rotating disk in a viscous fluid, *Phys. Fluids* **22**, 2267–2269.

Watson, P., Banks, W. H. H., Zaturska, M. B. and Drazin, P. G. 1991 Laminar channel flow driven by accelerating walls, *Eur. J. Appl. Math.* **2**, 359–385.

Weidman, P. D. and Mahalingam, S. 1997 Axisymmetric stagnation-point flow impinging on a transversely oscillating plate with suction, *J. Eng. Math.* **31**, 305–318.

Weidman, P. D. and Putkaradze, V. 2003 Axisymmetric stagnation flow obliquely impinging on a circular cylinder, *Eur. J. Mech. B, Fluids* **22**, 123–131.

Weidman, P. D. and Putkaradze, V. 2005 Erratum: 'Axisymmetric stagnation flow obliquely impinging on a circular cylinder', *Eur. J. Mech. B, Fluids* **24**, 788–790.

Weinbaum, S. and O'Brien, V. 1967 Exact Navier–Stokes solutions including swirl and cross flow, *Phys. Fluids* **10**, 1438–1447.

Weinbaum, S., Lawrence, C. J. and Kuang, Y. 1985 The inertial draining of a thin fluid layer between parallel plates with a constant normal force. Part 1. Analytical solutions; inviscid and small but finite-Reynolds-number limits, *J. Fluid Mech.* **156**, 463–477.

White, F. M. 1962 Laminar flow in a uniformly porous tube, *J. Appl. Mech.* **29**, 201–204.

Whitham, G. B. 1963 The Navier–Stokes equations of motion, *Laminar Boundary Layers*, ed. L. Rosenhead, Oxford: Clarendon Press, pp. 114–162.

Williams III, J. C. 1968 Nonsteady stagnation point flow, *AIAA J.* **6**, 2417–2419.

Wilson, L. O. and Schryer, N. L. 1978 Flow between a stationary and a rotating disk with suction, *J. Fluid Mech.* **85**, 479–496.

Womersley, J. R. 1955a Method for the calculation of velocity, rate of flow and viscous drag in arteries when the pressure gradient is known, *J. Physiol.* **127**, 553–563.

Womersley, J. R. 1955b Oscillatory motion of a viscous liquid in a thin-walled elastic tube: I. The linear approximation for long waves, *Philos. Mag.* (7) **46**, 199–221.

Wuest, W. 1952 Grenzschichten an zylindrischen Körpern mit nichtstationärer Querbewegung, *Z. Angew. Math. Mech.* **32**, 172–178.

Wybrow, M. F. and Riley, N. 1995 The flow induced by the torsional oscillations of an elliptic cylinder, *J. Fluid Mech.* **290**, 279–298.

Yang, K. T. 1958 Unsteady laminar boundary layers in an incompressible stagnation flow, *J. Appl. Mech.* **25**, 421–427.

Yatseyev, V. I. 1950 On a class of exact solutions of the equations of motion of a viscous fluid, *Zh. Eksp. Teor. Fiz.* **20**, 1031–1034 (in Russian).

Yih, C.-S., Wu, F., Garg, A. K. and Leibovich, S. 1982 Conical vortices: A class of exact solutions of the Navier–Stokes equations, *Phys. Fluids* **25**, 2147–2158.

Yuan, S. C. 1956 Further investigations of laminar flow in channels with porous walls, *J. Appl. Phys.* **27**, 267–269.

Zandbergen, P. J. 1980 New solutions of the Kármán problem for rotating flows, *Lecture Notes in Mathematics*, vol. **771**, Berlin: Springer-Verlag, pp. 563–581.

Zandbergen, P. J. and Dijkstra, D. 1977 Non-unique solutions of the Navier–Stokes equations for the Kármán swirling flow, *J. Eng. Math.* **11**, 167–188.

Zandbergen, P. J. and Dijkstra, D. 1987 Von Kármán swirling flows, *Annu. Rev. Fluid Mech.* **19**, 465–491.

Zaturska, M. B., Drazin, P. G. and Banks, W. H. H. 1988 On the flow of a viscous fluid driven along a channel by suction at porous walls, *Fluid Dyn. Res.* **4**, 151–178.

Index

For EU product safety concerns, contact us at Calle de José Abascal, 56–1°,
28003 Madrid, Spain or eugpsr@cambridge.org.